ELECTROMECHANICAL DEVICES

THEORY, APPLICATIONS, AND TROUBLESHOOTING

ELECTROMECHANICAL DEVICES

THEORY, APPLICATIONS, AND TROUBLESHOOTING

RICHARD A. HONEYCUTT

Electronics Instructor
Davidson County Community College
Lexington, North Carolina

PRENTICE-HALL, INC., Englewood Cliffs, New Jersey 07632

Library of Congress Cataloging in Publication Data

Honeycutt, Richard A.
   Electromechanical devices.

   Includes index.
   1. Electromechanical devices. 2. Electric
motors. 3. Electric generators. I. Title.
TK145.H743   1986   621.31'042   85-6356
ISBN 0-13-250077-9

*Editorial/production supervision and
   interior design:* Ellen Denning
Cover design: 20/20 Services, Inc.
Manufacturing buyer: Gordon Osbourne
Cover photograph courtesy of Century Electric, Inc.

Constructions shown for Figures 6-7 (b), 6-11 (a + b), and
8-6 (b) are covered by patents or applications for patents
and publication of these figures does not imply that these
devices or constructions are public property available for
unrestricted use. Courtesy of General Electric Co.

Printed in the United States of America

10 9 8 7 6 5 4 3 2 1

ISBN 0-13-250077-9 01

PRENTICE-HALL INTERNATIONAL (UK) LIMITED, *London*
PRENTICE-HALL OF AUSTRALIA PTY. LIMITED, *Sydney*
PRENTICE-HALL CANADA INC., *Toronto*
PRENTICE-HALL HISPANOAMERICANA, S.A., *Mexico*
PRENTICE-HALL OF INDIA PRIVATE LIMITED, *New Delhi*
PRENTICE-HALL OF JAPAN, INC., *Tokyo*
PRENTICE-HALL OF SOUTHEAST ASIA PTE. LTD., *Singapore*
EDITORA PRENTICE-HALL DO BRASIL, LTDA., *Rio de Janeiro*
WHITEHALL BOOKS LIMITED, *Wellington, New Zealand*

To Dad,
      who taught me to "use my head,"
      nurtured my early interest in things electrical,
      led me by example to writing and teaching,
      and also taught me about life;

and to Mom,
      who still does . . .

CONTENTS

5   DC MOTORS   111

6   AC MOTORS   134

Appendices

PREFACE

This book was written to provide a working understanding of electric motors, relays, and related devices. It presents the operating principles of these devices in a simple, nonmathematical fashion. The object is to enable an electronic technician to specify replacement devices intelligently, and to perform the many simple repairs that electrical machines occasionally require. The book also includes a discussion of linear-motion devices, not only because they properly belong to the family treated in this book, but also because they are often not treated at all elsewhere.

The first chapter covers such introductory ac and dc theory and circuits as are needed for an understanding of the material to follow. It makes the book suitable for readers with no electrical background, for students of industrial maintenance, appliance repair, refrigeration, and the like. It can also serve as a refresher for those who have already covered similar material elsewhere. Readers with some background in electrical and electronics technology may want to read the introductory chapter, skip Chapter 1, skim Chapter 2 (magnetism), and then begin in earnest with Chapter 3.

I am grateful for the help and support of my (extended!) family during the preparation of this book, especially manuscript copyists Alyson, April, and Erin. And the Without-Whom-This-Book-Would-Still-Be-In-Preparation Award goes jointly to Peggy and Betty Jane for their immense help in typing the manuscript.

ACKNOWLEDGMENTS

I would like to gratefully acknowledge the help of the following companies in supplying the indicated material, and giving permission for its reproduction:

Airpax Corporation, information and illustrations on stepper motors

Allen-Bradley Co., photographs of resistors

Bodine Electric Co., photographs of motors and motor parts; also permission to reproduce material from *Small Motor, Gearmotor, and Control Handbook*, © 1978 by Bodine Electric Co.

C&K Components, Inc., photographs of switches

Century Electric, Inc.—especially Mr. Gordon Quigley—information and photographs on solid-state starting switches

Delco-Remy Division of General Motors Corporation, photographs and materials on alternators and voltage regulators

General Electric Co., photographs of motors and motor parts

IMC Magnetics Corporation, photographs of solenoids

Switchcraft Inc., photographs of connectors

And especially to Mr. Dean Charlton, for many of the other photographs

RICHARD A. HONEYCUTT

ELECTROMECHANICAL DEVICES

THEORY, APPLICATIONS, AND TROUBLESHOOTING

GETTING STARTED

Interesting, isn't it? We have all been told more times than we care to count that: "We live in an era of microelectronics. In 72 cubic inches your car stereo combines functions that would have taken up half a bookcase 10 years ago. Microcomputers the size of your hand can perform control functions that . . ." and so on and so forth. You've heard it. But even in an age of microelectronics, something still has to control the machines, generate the power, and do the work. These devices—relays, generators, motors (rotary and linear)—are the subject of our present study. In the chapters that follow we will do a number of things. First, we will review the electrical circuits and components that are used in connection with electromechanical devices. Second, we will study enough of the theory of magnetism so that you can understand how relays, generators, and motors work. Then we will get into the meat of the material: the devices themselves.

Relays of many varieties are found in all kinds of locations: automobiles, office machines, appliances, test equipment, stereo amplifiers and receivers, and all sorts of industrial controllers. We will discuss the many varieties of relays, how they work, and what circuits are used to operate them.

Dc and ac generators are used (obviously!) for generating power, not only in the large electric utilities' generating plants, but in the standby power systems used in hospitals, telephone offices, radio stations, and industrial plants. Generators are used in autos, trucks, tractors, and airplanes. Instruments such as tachometers frequently use small generators whose output is proportional to the rotation speed. Your cassette tape deck and turntable are likely to contain generators that are part of the speed-control system.

We will see what different types of generators there are and how they operate.

Motors are everywhere; what more need we say? Even among non-technicians, rare is the person who has never needed to replace a motor. But for the technician, replacement is often more involved than just "picking up a new motor." Is it ac or dc? What voltage? How many phases? What horsepower? What rpm? For a hard-to-start or an easy-to-start load? What type of casing? To answer these questions, the technician must know a good bit about how motors operate. In the process of discussing motor operation, we will see what sorts of special circuits are used to start, stop, and reverse them and to control their speed. We will also discuss several devices that are closely akin to motors.

And when they break. . . . What do you do when a relay or generator or motor stops working? The usual answer is replacement or, for a motor, rewinding. But the usual answer is a wasteful one. Probably 95% of all defective relays can be repaired quite simply. Most defective generators simply require brush or bearing replacement. Roughly 50% of all defective power tools and 80% of all defective single-phase ac motors can be repaired by a competent technician without rewinding or special tools. Sometimes the circuitry associated with the relay, generator, or motor is at fault rather than the device itself. We will see how to troubleshoot and repair these devices and at what point a repair should be farmed out to a shop with specialized equipment.

Although it may seem a far jump from industrial motors to chart-recorder pen drives and loudspeakers, the operating principles are really very closely related. The average technician may never be called on to actually repair a linear-motion electromechanical device, but if the technician understands how they operate, he or she will be able both to connect them correctly in an installation and to talk intelligently when specifying a replacement unit.

The study of electromechanical devices, then, is quite useful, whether you are in an industrial environment, research, or even consumer equipment repair. And since the average home has over a hundred of these devices, you may also be able to save a little money on repairs—even in an era of microelectronics.

1

CIRCUITS
AND COMPONENTS

A thorough discussion of basic ac and dc electrical circuits is not really within the scope of this book. However, just in case you are a bit rusty on (or unfamiliar with) basic circuit concepts, we'll begin with a review of this material. First we discuss dc circuits; we look at ac circuits beginning on page 14. If you decide to skip these sections, it would still be a good idea to read the material on components and power-supply circuits starting on page 22.

SERIES RESISTIVE CIRCUITS

A series circuit is one that has only one path for electric current. If the circuit has no capacitors or inductors—only resistors and wire—it is said to be a *resistive circuit.* Since electrons are neither produced nor destroyed in any circuit, *the current is the same at all points in any complete resistive series circuit.* (The only device that is allowed to "pile up" electrons is a capacitor, and a series circuit containing a capacitor is not a resistive circuit.) The total voltage applied to a resistive series circuit is shared among the various devices in the circuit according to their resistances. The amount of voltage appearing across a device is related to the device's resistance by Ohm's law: $E = IR$, where E = voltage, I = current, and R = resistance. The total of all the individual voltage drops in a series dc circuit must add up to the total applied voltage. This obvious-sounding fact is called *Kirchhoff's voltage law.* Figure 1-1 gives an example that we can use to show how all these things work together.

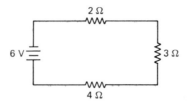

Figure 1-1   Voltage drops in series dc circuits.

$$I_{total} = \frac{E_{total}}{R_{total}} \times \frac{6 \text{ V}}{2 \text{ }\Omega + 3 \text{ }\Omega + 4 \text{ }\Omega} = \frac{2}{3} \text{ A}$$

$$E_{2\Omega} = I_{total} \times R_2 = \frac{2}{3} \text{ A} \times 2 \text{ }\Omega = 1\frac{1}{3} \text{ V}$$

$$E_{3\Omega} = I_{total} \times R_3 = \frac{2}{3} \text{ A} \times 3 \text{ }\Omega = 2 \text{ V}$$

$$E_{4\Omega} = I_{total} \times R_4 = \frac{2}{3} \text{ A} \times 4 \text{ }\Omega = 2\frac{2}{3} \text{ V}$$

Notice that the total current is found first; this equals total voltage divided by total resistance. The total resistance is just the sum of the individual resistances. Next, the voltage drops across the individual resistors are calculated. Just as a check, let's see if our results agree with Kirchhoff's voltage law:

$$E_T = E_{2\Omega} + E_{3\Omega} + E_{4\Omega}$$

$$= 1\frac{1}{3} \text{ V} + 2 \text{ V} + 2\frac{2}{3} \text{ V}$$

$$= 6 \text{ V}$$

Since applied voltage equals the sum of the individual voltage drops, our calculations are correct.

It is not necessary that wires be used for a complete series circuit. Often a part of the current path consists of the metal chassis of a piece of equipment. Circuits that are completed in this way are called *ground-return circuits* because the chassis is usually *circuit common*, or *ground*. An example of such a circuit is shown in Fig. 1-2.

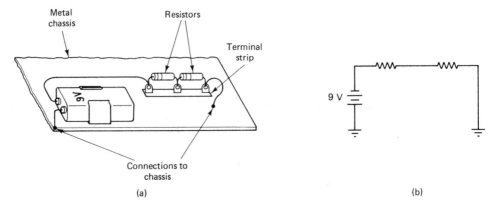

(a)   (b)

Figure 1-2   Ground-return series circuit: (a) pictorial; (b) schematic.

When a technician wants to know the voltage drop across a certain component in a ground-return circuit, there are two ways to measure it. The first is simply to connect a voltmeter across the component. The second is to measure the voltage from each side of the component to ground, then subtract. Both methods are illustrated in Fig. 1-3. Although the second method sounds like extra trouble, it is actually the more common way of doing things. The reason is that in a schematic diagram the voltage-to-ground at each point can be specified much more easily than the voltage from each point to each other point. *Thus, whenever a voltage is specified in a circuit diagram, it is assumed to be the voltage from that point to the circuit ground (usually chassis) unless indicated otherwise.*

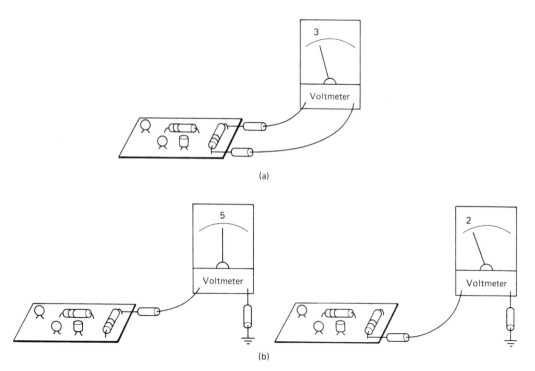

Figure 1-3 Measurement of voltage drop: (a) direct measurement; (b) indirect measurement.

Notice that a voltmeter is always connected *across* the points to be measured. Current measurement cannot be done in the same way. To measure current, the ammeter must be connected *in series with* the circuit. In other words, the circuit must be opened and the ammeter inserted into it so that the current flows through the meter. Figure 1-4 illustrates this difference.

When two resistors are connected in series, the applied voltage is divided according to the ratio of the resistances. Thus such a circuit is called a

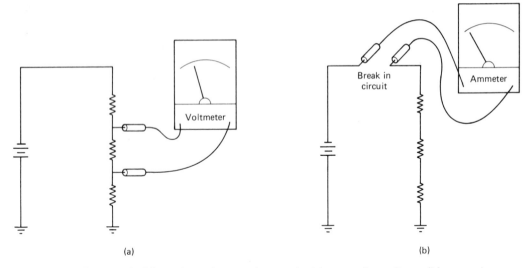

Figure 1-4 Measuring voltage and current: (a) measuring voltage; (b) measuring current.

voltage-divider circuit. The voltage at the junction of the two resistors is given by the *voltage-division rule*:

$$E_J = E_T \left(\frac{R_2}{R_1 + R_2} \right)$$

If a group of resistors are connected in series, the resistances above and below the junction (J) point can be added to make up equivalent R_1 and R_2 resistances. Figure 1-5a and b illustrate voltage division. Part (c) of the figure shows a common use of the voltage-division principle. If it is desired to apply less voltage to a load than is supplied by the battery (or other power-supply device), part of the voltage can be dropped across a series "dropping" resistor. The voltage applied to the load is reduced, as given by the voltage-division rule:

$$E_J = E_T \left(\frac{R_\text{load}}{R_\text{dropping} + R_\text{load}} \right)$$

The load current is reduced correspondingly.

The power dissipated in a component or circuit is equal to the product of voltage and current: $P = EI$. It is important to remember that the values of voltage and current used in the formula above are the voltage across and current in the component or circuit in question. In our earlier example using a 6-V battery and a 2-Ω, a 3-Ω, and a 4-Ω resistor, the total power would be

$$P = E \times I = 6 \text{ V} \times \text{2/3 A} = 4 \text{ W}$$

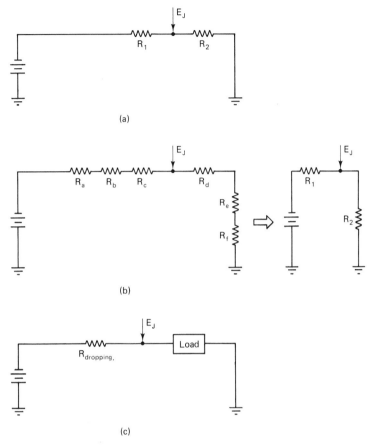

Figure 1-5 Voltage division: (a) simple circuit; (b) more resistors; (c) "dropping" resistor to limit load voltage or current.

Each individual resistor's power would be the current multiplied by the voltage across that resistor:

$$P_2 = E_2 \times I_T = 1^1/_3 \text{ V} \times {}^2/_3 \text{ A} = {}^8/_9 \text{ W}$$

$$P_3 = E_3 \times I_T = 2 \text{ V} \times {}^2/_3 \text{ A} = 1^1/_3 \text{ W}$$

$$P_4 = E_4 \times I_T = 2^2/_3 \text{ V} \times {}^2/_3 \text{ A} = 1^7/_9 \text{ W}$$

Adding, we get a total power of 4 W, which is just what we calculated before.

Many series circuits use power-supply devices other than batteries. These may include generators, solar cells, heat cells, or other devices. To indicate a general dc voltage source, a schematic symbol was adopted, consisting of a circle with a "+" sign on one side to indicate polarity (see Fig. 1-6a). Sometimes several of these devices are connected in the same series circuit. If they are connected with the same polarity, they are said to be *series-aiding*

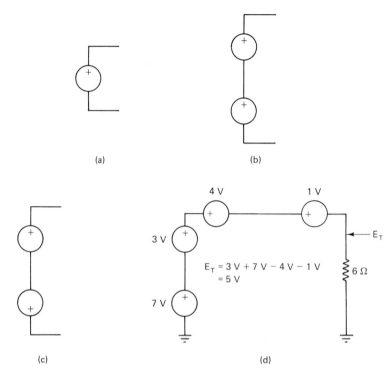

Figure 1-6 Series-connected voltage sources: (a) symbol for voltage source; (b) series-aiding sources; (c) series-opposing sources; (d) combined sources.

(see Fig. 1-6b). If they are connected with opposite polarity, they are *series-opposing* (see Fig. 1-6c). The total voltage of any group of series-connected voltage sources is equal to the algebraic sum of the individual voltages. Another way of saying this is: To obtain the total voltage of a group of series-connected voltage sources, just add all the series-aiding voltages and subtract the series-opposing ones. Thus the current in the 6-Ω resistor in Fig. 1-6d would be

$$I_T = \frac{E_T}{R} = \frac{5\text{ V}}{6\text{ Ω}} = \frac{5}{6}\text{ A}$$

You may be wondering now why we are seeing so many equations. After all, this book is supposed to teach you to *repair* equipment, for heaven's sake, not calculate its operation! The reason is that there is always a certain amount of information that a technician needs that is not given on a schematic diagram. In fact, there may be no schematic available for much of the equipment you will repair. You may have to calculate replacement values for unmarked components or for components whose markings have been de-

stroyed by overheating. The basic calculation methods shown in this chapter
will help you whenever you face these situations.

PARALLEL RESISTIVE CIRCUITS

Many electric circuits offer more than one path for current. Such circuits are
called *parallel circuits.* In a parallel circuit, each current path is called a
branch. Since all branches are fed from the same two points, the voltage
across each branch is the same. The current in each branch depends on the
resistance in that branch, and the total current equals the sum of the branch
currents. This is *Kirchhoff's current law.*

By Ohm's law (see Fig. 1-7),

$$I_{2\Omega} = \frac{E_T}{2\ \Omega} = 3 \text{ A}$$

$$I_{3\Omega} = \frac{E_T}{3\ \Omega} = 2 \text{ A}$$

$$I_{4\Omega} = \frac{E_T}{4\ \Omega} = 1.5 \text{ A}$$

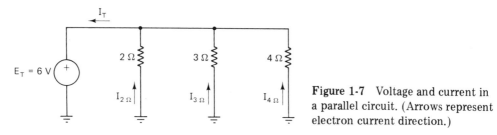

Figure 1-7 Voltage and current in
a parallel circuit. (Arrows represent
electron current direction.)

Since the total current must equal the sum of the branch currents, we
can calculate a total equivalent resistance for this parallel circuit:

$$I_T = I_{2\Omega} + I_{3\Omega} + I_{4\Omega}$$

$$= 3 \text{ A} + 2 \text{ A} + 1.5 \text{ A} = 6.5 \text{ A}$$

$$R_T = \frac{E_T}{I_T} = \frac{6 \text{ V}}{6.5 \text{ A}} = 0.923 \ \Omega$$

This same value can be calculated more directly by the formula

$$\frac{1}{R_T} = \frac{1}{R_1} + \frac{1}{R_2} + \frac{1}{R_3} + \dots + \frac{1}{R_n}$$

Thus

$$\frac{1}{R_T} = \frac{1}{2\,\Omega} + \frac{1}{3\,\Omega} + \frac{1}{4\,\Omega} = 0.5\,\mho + 0.33\,\mho + 0.25\,\mho = 1.08\,\mho$$

and

$$R_T = \frac{1}{1.08\,\mho} = 0.923\,\Omega$$

($\mho$ is the symbol for $1/\Omega$, which is conductance, and is measured in sie-mens.) Notice that the total equivalent resistance is less than any of the indi-vidual resistances. This is always true in a parallel circuit. There are some shortcuts for special cases of parallel resistors:

1. Only two resistors: $R_T = \dfrac{R_1 R_2}{R_1 + R_2}$

2. Two identical resistors: $R_T = {}^1\!/_2\,R$

3. Three identical resistors: $R_T = {}^1\!/_3\,R$

4. n identical resistors: $R_T = \dfrac{1}{n}\,R$

Power dissipation in the branches of a parallel circuit is calculated in exactly the same way as for a series circuit: The current in the branch is mul-tiplied by the voltage across the branch. Since the current is greater for smaller resistances, the greatest amount of power is dissipated by the lowest-resistance branch. Total power, of course, equals the sum of the branch powers.

If identical voltage sources are connected in parallel, the output voltage remains the same but the current capability of the several sources adds. Thus, if three 6-V batteries are connected in parallel and each battery is rated to provide a 1-A current over an 8-hour discharge time, the total cur-rent capability is 3 A. If the voltage sources having different voltages are con-nected in parallel, the resulting voltage will be difficult to predict.

To explain why, we first need to discuss *internal resistance*. Since no real voltage source is perfect, there is always a certain amount of resistance in the conductors that make up the source. For example, a zinc-carbon bat-tery cell produces 1.55 V. As a battery ages, however, its output voltage drops. This is not because of a change in the chemical action that produces the electricity, but because of increasing internal resistance. In fact, if a weak zinc-carbon battery's voltage is measured in some way that draws almost no current, it will read nearly 1.5 V. When the weak battery is asked to produce 1 A or so to light a flashlight bulb, its internal resistance may drop so much voltage that the battery only produces 0.5 V or less at its terminals.

Figure 1-8a illustrates a symbol that can be used to represent a real volt-age source, that is, one whose internal components and conductors have some resistance. The output of such a voltage source is

$$E_{terminal} = E_{source} - I \times R_{internal}$$

For example, if E_{source} (the no-load voltage) equals 10 V and $R_{internal}$ equals 0.1 Ω then:

For $I = 0$: $E_{terminal} = 10 \text{ V} - 0 \text{ A} \times 0.1 \text{ Ω} = 10 \text{ V}$

For $I = 1 \text{ A}$: $E_{terminal} = 10 \text{ V} - 1 \text{ A} \times 0.1 \text{ Ω} = 9.9 \text{ V}$

For $I = 5 \text{ A}$: $E_{terminal} = 10 \text{ V} - 5 \text{ A} \times 0.1 \text{ Ω} = 9.5 \text{ V}$

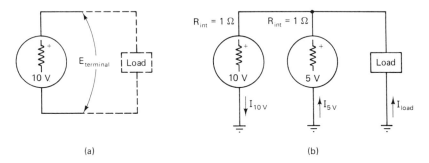

(a)   (b)

Figure 1-8 Parallel-connected nonidentical voltage sources: (a) voltage source with internal resistance; (b) parallel connection of voltage sources.

If we connect nonidentical voltage sources in parallel, the terminal voltages will obviously have to have the same value, since it is not possible to have two different voltages at the same place at the same time! This means that the higher-voltage source will have to pump current backward through the lower-voltage source so that the voltage drops across their internal resistances make the outputs equal. This is much easier to draw than to describe (see Fig. 1-8b). Notice the direction of current in the 5-V source. Let's calculate the terminal voltages for the two sources:

For the 10-V source:

$$E_{terminal} = 10 \text{ V} - (I_{load} + I_{5V}) \times 1 \text{ Ω}$$

For the 5-V source:

$$E_{terminal} = 5 \text{ V} + (I_{5V} \times 1 \text{ Ω})$$

The backward flow of current in the 5-V source causes the voltage drop in its internal resistance to *add* to the source voltage. Since the terminal voltages must be equal, the two expressions above must be equal:

$$10 \text{ V} - (I_{load} + I_{5V}) \times 1 \text{ Ω} = 5 \text{ V} + (I_{5V} \times 1 \text{ Ω})$$

$$10 \text{ V} - I_{load} \times 1 \text{ Ω} - I_{5V} \times 1 \text{ Ω} = 5 \text{ V} + (I_{5V} \times 1 \text{ Ω})$$

Since $I_{load} = 3$ A:

$$10 \text{ V} - 3 \text{ A} \times 1 \text{ }\Omega - I_{5V} \times 1 \text{ }\Omega = 5 \text{ V} + I_{5V} \times 1 \text{ }\Omega$$

or

$$10 \text{ V} - 5 \text{ V} - 3 \text{ V} = 2(I_{5V} \times 1 \text{ }\Omega)$$

$$2 \text{ V} = 2(I_{5V} \times 1 \text{ }\Omega)$$

$$1 \text{ V} = I_{5V} \times 1$$

$$I_{5V} = \frac{1 \text{ V}}{1} = 1 \text{ A}$$

Thus we see that the 10-V source forces 1 A backward through the 5-V source, making a total draw of 4 A from the 10-V source.

If the internal resistance were different, this value and the output voltage would be different. The important thing to notice is that the extra drain on the high-voltage source may shorten its life or even cause so much heat that one or both sources will burn out. This fact is very important in cases where a switchover from one power supply to another must occur without an interruption, so that the two supplies are momentarily connected in parallel. This situation will be discussed in Chapter 4.

COMBINATION RESISTIVE CIRCUITS

If one part of a circuit is series and another part is parallel, the whole circuit can be called a *combination*, a *series-parallel*, or a *complex* circuit. The term "complex" raises students' blood pressure too much, and "series-parallel" takes too long to write, so we'll use the term "combination circuits." There are two families of combination circuits. One is a series of parallel circuits (Fig. 1-9a), and the other is a parallel arrangement of series circuits (Fig. 1-9b). Circuits of the first type can be dealt with by finding the equivalent resistance of each parallel group, then adding those equivalent resistances:

$$R_{equiv \, 1} = \frac{2 \text{ }\Omega \times 3 \text{ }\Omega}{2 \text{ }\Omega + 3 \text{ }\Omega} = 1.2 \text{ }\Omega$$

$$R_{equiv \, 2} = \frac{4 \text{ }\Omega \times 5 \text{ }\Omega}{4 \text{ }\Omega + 5 \text{ }\Omega} = 2.22 \text{ }\Omega$$

$$R_T = R_{equiv \, 1} + R_{equiv \, 2} = 1.2 \text{ }\Omega + 2.22 \text{ }\Omega$$

$$= 3.42 \text{ }\Omega$$

The total current is

$$I_T = \frac{E_T}{R_T} = \frac{10\text{ V}}{3.42\ \Omega} = 2.92\text{ A}$$

The voltage across $R_{\text{equiv 2}}$ can be found using the voltage-division rule:

$$E_J = E_T\left(\frac{R_{\text{equiv 2}}}{R_T}\right) = 10\text{ V}\left(\frac{2.22\ \Omega}{3.42\ \Omega}\right) = 6.5\text{ V}$$

The current through each resistor can then be found by dividing the voltage across that resistor by the resistance. You can do that now; it's a good exercise!

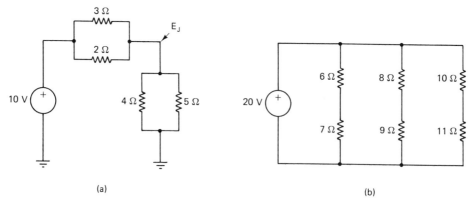

(a)   (b)

Figure 1-9 Combination circuits: (a) series-connected parallel combinations; (b) parallel-connected series combinations.

Circuits like the one in Fig. 1-9b can be solved by adding the series resistors, then finding the equivalent resistance of the resulting parallel circuit. A simpler way is to find the individual currents and voltages first:

$$\text{Branch 1: } I_1 = \frac{E_T}{R_1} = \frac{20\text{ V}}{6\ \Omega + 7\ \Omega} = 1.54\text{ A}$$

$$\text{Branch 2: } I_2 = \frac{E_T}{R_2} = \frac{20\text{ V}}{8\ \Omega + 9\ \Omega} = 1.18\text{ A}$$

$$\text{Branch 3: } I_3 = \frac{E_T}{R_3} = \frac{20\text{ V}}{10\ \Omega + 11\ \Omega} = 0.95\text{ A}$$

$$I_T = I_1 + I_2 + I_3 = 1.54\text{ A} + 1.18\text{ A} + 0.95\text{ A}$$
$$= 3.67\text{ A}$$

Then, for the voltages across the individual resistors:

$$\text{Branch 1:} \quad E_{6\Omega} = I_1 R_{6\Omega} = 1.54 \text{ A} \times 6 \text{ } \Omega = 9.24 \text{ V}$$

$$E_{7\Omega} = \text{the rest of the 20 V} = 10.76 \text{ V}$$

You can calculate the values for branches 2 and 3. Finally, if you want to know R_T:

$$R_T = \frac{E_T}{I_T} = \frac{20 \text{ V}}{3.67 \text{ A}} = 5.45 \text{ } \Omega$$

AC CIRCUITS WITH RESISTANCE AND INDUCTANCE (RL CIRCUITS)

Resistors are very tolerant creatures. They behave the same for either ac or dc. Not so, however, with inductors (coils) and capacitors. An inductor causes a phase shift in a circuit; it causes the voltage wave to lead the current wave by 90°. This has the peculiar effect that if an inductor and a resistor are connected in series in an ac circuit, and each has a 1-V drop across it, the total drop is not 2 V! Let's see why.

First, we need to discuss a concept called *reactance. Reactance is the ratio of voltage to current in a resistanceless ac circuit.* Reactance is measured in ohms. This sounds quite a bit like resistance, but there is one big difference. Resistance converts electrical energy to heat (i.e., it dissipates power); reactance does not. The reactance of an inductor depends on its inductance and the frequency of the ac:

$$X_L = 2\pi f L$$

where X_L = inductive reactance, Ω

f = frequency, in hertz, Hz

L = inductance, in henries, H

For example, a 1-H inductor operating with 60 Hz ac would have a reactance of

$$X_L = 6.28 \times 60 \text{ Hz} \times 1 \text{ H}$$

$$= 376.8 \text{ } \Omega \quad \text{or roughly} \quad 377 \text{ } \Omega$$

Notice that inductance is a characteristic of the coil; inductive reactance describes how the coil behaves *with ac of a certain frequency.*

Next, we need to introduce phase diagrams. These are ways of drawing pictures of the phase relations in an ac circuit. Figure 1-10 shows several examples. Each line on the diagram is called a *phasor. A phasor is a quantity that has a magnitude (or amount) and a phase angle.* For example, since the voltage in an inductive circuit leads the current by 90°, the inductive reactance we calculated in the example above would be represented by the phasor: $377\,\Omega\underline{/90°}$. There is no phase shift in a resistive circuit, so a 377-Ω resistor would be represented by the phasor: $377\,\Omega\underline{/0°}$. On the diagram, the length of the line represents the magnitude of the phasor, and the angle from horizontal represents the phase angle. Notice that in Fig. 1-10c a diagram is shown for a circuit containing a resistor and an inductor. The combined effect of the resistance and the inductive reactance is shown by the line that forms the third side of the triangle. This line is called the *resultant*. It represents the impedance of the circuit. *Impedance is the ratio of voltage to current in an ac circuit that has both resistance and reactance.* Table 1-1 makes all this clearer.

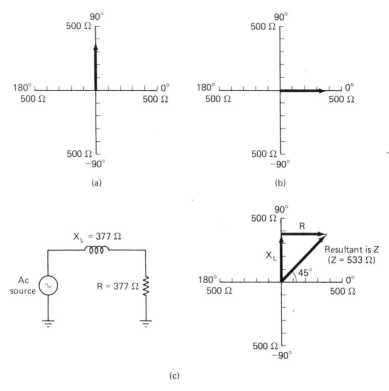

(a)

(b)

(c)

Figure 1-10 Phase diagrams for *RL* circuits: (a) $X_L = 377\ \Omega$; (b) $R = 377\ \Omega$; (c) X_L and X_R in series.

TABLE 1-1 Resistance, Reactance, and Impedance

Quantity	Measured in:	Phase angle	Dissipates power?
Resistance	Ohms	$0°$	Yes
Reactance	Ohms	$\pm90°$	No
Impedance	Ohms	Any angle	Resistive part only

The impedance can be scaled from a phasor diagram with a ruler and a protractor, or it can be calculated. The magnitude is calculated by means of the Pythagorean theorem, which gives the length of the hypoteneuse of a right triangle if the lengths of the other sides are known. In its original form it is

$$(\text{hypoteneuse})^2 = (\text{opposite})^2 + (\text{adjacent})^2$$

where the hypoteneuse, opposite, and adjacent are the names of the three sides of a right triangle. For our purposes, we can simply write

$$Z^2 = X_L^2 + R^2$$

where Z = impedance

X_L = inductive reactance

R = resistance

Or we can rewrite the equation as

$$Z = \sqrt{X_L^2 + R^2}$$

Since Z is a phasor quantity, and we are discussing only its magnitude, we should write

$$|Z| = \sqrt{|X_L|^2 + R^2}$$

The vertical lines enclosing the Z and the X_L indicate "magnitude of impedance" and "magnitude of inductive reactance."

In our earlier example with $|X_L| = 377\ \Omega$ and $R = 377\ \Omega$, $|Z|$ would be

$$|Z| = \sqrt{|X_L|^2 + R^2}\ \Omega$$
$$= \sqrt{377^2 + 377^2}\ \Omega$$
$$= \sqrt{142{,}129 + 142{,}129}\ \Omega$$
$$= \sqrt{284{,}248}\ \Omega = 533\ \Omega$$

Now how about the phase angle? Well, that's easier:

$$\theta_Z = \tan^{-1}\left(\frac{|X_L|}{R}\right)$$

where θ_Z is the phase angle of the impedance (θ is the Greek lowercase letter theta—"thay-ta") and $\tan^{-1}$ means "inverse tangent." The inverse tangent of a number is the angle of which that number is the tangent. If you're using a table like the one in Appendix C, just look up the number given by $|X_L|/R$ in the table and find the angle it corresponds to. If you're using an electronic calculator, look in your instruction manual for the correct process. In our example,

$$\frac{|X_L|}{R} = \frac{377\ \Omega}{377\ \Omega} = 1$$

so we find that $\theta_Z = 45°$. The complete impedance of our circuit is then $533\ \Omega\underline{/45°}$.

The remaining question you may have is: "What if you have more than one inductor?" The answer is that inductances combine just as resistances do. Inductances in series add, and inductances in parallel combine according to

$$\frac{1}{L_{\text{total}}} = \frac{1}{L_1} + \frac{1}{L_2} + \frac{1}{L_3} + \ldots + \frac{1}{L_n}$$

Inductive reactances combine in the same way.

AC CIRCUITS WITH RESISTANCE AND CAPACITANCE (*RC* CIRCUITS)

A student once told me, "Capacitors and inductors are just alike, except they're different!" Actually, he was right. They are alike in that they both have reactances and cause 90° phase shifts. They are alike in that neither dissipates power. They are different in that the phase shifts are in opposite directions, and the reactances are figured differently. Also, multiple capacitors combine differently from multiple inductors. But one thing at a time!

A capacitor stores electrical energy. Once it is charged, it acts as an open circuit to dc. But an ac series circuit containing a capacitor is a complete circuit. The capacitor will have a voltage drop across it, because it has a capacitive reactance, given by

$$|X_C| = \frac{1}{2\pi f C}$$

where $|X_C|$ = magnitude of capacitive reactance, Ω

f = frequency of ac, Hz

C = capacitance, in farads, F

For example, a 7.04-μF capacitor in a 60-Hz circuit has a reactance of

$$|X_C| = \frac{1}{6.28 \times 60\ \text{Hz} \times (7.04 \times 10^{-6})\text{F}} \simeq 377\ \Omega$$

In a capacitive circuit the voltage lags the current by 90°, so the phase angle of capacitive reactance is $-90°$. Figure 1-11 shows phase diagrams for a circuit containing resistance and capacitance. Notice that they look just like the ones given in Fig. 1-10 for *RL* circuits, except that the phasor points down instead of up. Consequently, the resultant points to the lower right $(-45°)$ rather than the upper right $(45°)$.

The magnitude and phase angle of the impedance are figured just as for RL circuits.

$$|Z| = \sqrt{|X_C|^2 + R^2}$$
$$= \sqrt{377\ \Omega + 377\ \Omega} = 533\ \Omega$$

$$\theta_Z = \tan^{-1}\left(\frac{|X_C|}{R}\right) = -45°$$

We know it is $-45°$ rather than $+45°$ because the reactance is capacitive.

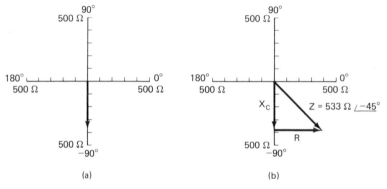

(a)   (b)

Figure 1-11   Phase diagrams for RC circuits: (a) $X_C = 377\ \Omega$; (b) X_C and X_R in series.

Well, then, do capacitors combine in the same way that resistors do? No! The total capacitance of capacitors in *parallel* is equal to the sum of the individual capacitances:

$$C_T = C_1 + C_2 + C_3 + \ldots + C_n$$

The total capacitance of capacitors in *series* is calculated like the total resistance of resistors in parallel:

$$\frac{1}{C_T} = \frac{1}{C_1} + \frac{1}{C_2} + \frac{1}{C_3} + \ldots + \frac{1}{C_n}$$

Capacitive reactances, though, *do* behave like resistances; they add in series and their inverses are added when they are in parallel. The paradox is cleared up if you remember that inductive reactance varies directly with inductance, but capacitive reactance varies inversely with capacitance. Thus L, X_L, and X_C combine like resistors, but C combines inversely.

RLC **CIRCUITS**

Ah, yes, you were right; all this does tie together. And much more simply
than you may have thought. First, let's review what happens in a reactive circuit. Since the current is not allowed to bunch up anywhere, current is in
phase at all points in a series circuit. The phase shift that occurs causes the
voltage to lead or lag the current, (i.e., the voltage at all points is not necessarily in phase). Now what happens if you shift the voltage ahead by, say,
73°, then later shift it behind by 21°? The net effect is that voltage will be
shifted ahead by something less than 73°. In other words, *the effects of inductive reactance and capacitive reactance cancel each other:*

$$|X_T| = |X_L| - |X_C|$$

If you get a negative number from this equation, it means that you will have
a negative phase angle:

$$\theta_Z = \tan^{-1}\left(\frac{|X_L| - |X_C|}{R}\right)$$

The impedance of an *RLC* circuit is

$$|Z| = \sqrt{(|X_L| - |X_C|)^2 + R^2}$$

Remember that the square of a negative number is a positive number, so
even if $|X_L| - |X_C|$ is negative, $(|X_L| - |X_C|)^2$ will be positive and will be
added to R^2. Figure 1-12 shows a phase diagram for the impedance of an
RLC circuit. For this circuit:

$$|X_L| = 2\pi f L = 6.28 \times 60 \text{ Hz} \times 0.15 \text{ H} = 56.5 \ \Omega \ \underline{/90°}$$

$$X_C = \frac{1}{2\pi f C} = \frac{1}{6.28 \times 60 \text{ Hz} \times (22 \times 10^{-6} \text{ F})}$$

$$= 120.6 \ \Omega \ \underline{/-90°}$$

$$|Z| = \sqrt{(|X_L| - |X_C|)^2 + R^2}$$

$$= \sqrt{(56.5 \ \Omega - 120.6 \ \Omega)^2 + 75 \ \Omega^2}$$

$$= \sqrt{4108 + 5625} \ \Omega = 98.8 \ \Omega$$

$$\theta_Z = \tan^{-1}\left(\frac{|X_L| - |X_C|}{R}\right)$$

$$= \tan^{-1}\left(\frac{56.5 \ \Omega - 120.6 \ \Omega}{75 \ \Omega}\right)$$

$$= \tan^{-1}(-0.855) = -40.5°$$

So

$$Z = 98.8 \ \Omega \ \underline{/-40.5°}$$

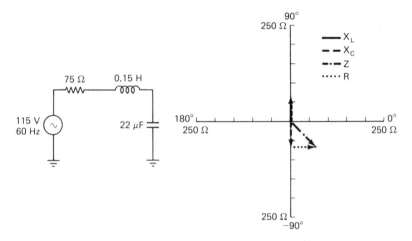

Figure 1-12   Solution for the impedance of an *RLC* circuit.

Well, now, what good is all this once we have found the answer? Just watch:

$$E = I \times Z \ (\text{"AC Ohm's law"}),$$

so we find the current in the circuit by:

$$I = \frac{E}{Z}$$

For example:

$$I = \frac{115 \ V}{98.8 \ \Omega} = 1.16 \ A$$

We can find the voltage drop across the inductor by

$$E_L = I \times X_L$$

For example:

$$E_L = 1.16 \ A \times 56.5 \Omega \underline{/90°}$$
$$= 65.5 \ V \underline{/90°}$$

We can find the voltage drop across the capacitor by

$$E_C = I \times X_C$$

For example:

$$E_C = 1.16 \ A \times 120.6 \Omega \underline{/-90°}$$
$$= 139.9 \ V \underline{/-90°}$$

And of course the voltage drop across the resistor is

$$E_R = I \times R$$

For example:

$$E_R = 1.16 \text{ A} \times 75 \text{ }\Omega = 87 \text{ V}$$

Notice that the sum of the voltage drops is 292.4 V. But we only applied 115 V, you say. Remember that capacitive action opposes inductive action? Let's see what happens if we account for that in the equations:

$$E_s = \sqrt{(|E_L| - |E_C|)^2 + E_R^2}$$
$$= \sqrt{(65.5 \text{ V} - 139.9 \text{ V})^2 + (87)^2}$$
$$= \sqrt{5535 + 7569} \text{ V} = 114.5 \text{ V}$$

where E_s is the supply voltage. We lost the other five-tenths of a volt by rounding off. Notice two things. First, the two reactive voltages tend to cancel each other. Second, the 90°-out-of-phase reactive and resistive voltages must be combined via the Pythagorean theorem, just like reactance and resistance.

Now how about the power? Total voltage = 115 V, and total current = 1.16 A. So apparently the power is

$$115 \text{ V} \times 1.16 \text{ A} = 133.4 \text{ W}$$

But only the resistor can dissipate power. Its voltage and current give

$$87 \text{ V} \times 1.16 \text{ A} = 100.9 \text{ W}$$

Which is correct? Actually, the 133.4-W figure is called the *apparent power*:

$$P_{\text{app}} = E_s \times I_s$$

The 100.9-W figure is the *true power*:

$$P_{\text{true}} = E_R \times I_R = I_R^2 \times R$$

Since only the true power is converted into work, and industrial companies are interested in having their equipment do as much work as possible per watt input, the relationship of true power to apparent power is important. This relationship is called the *power factor*:

$$\text{PF} = \frac{P_{\text{true}}}{P_{\text{app}}}$$

Obviously, the ideal power factor is 1. A low power factor usually indicates low-efficiency operation. The power factor can also be calculated from the phase angle of the impedance:

$$\text{PF} = \cos \theta_Z$$

So in our example, we find that

$$PF = \frac{P_{true}}{P_{app}} = \frac{100.9\ W}{133.4\ W} = 0.757$$

or

$$PF = \cos\theta_Z = \cos 40.5° = 0.760$$

Once again, the small inaccuracy was caused by rounding off.

One last question: What if the capacitive and inductive reactances were equal? The answer is that they would completely cancel and the impedance would be equal to the resistance with a 0° phase angle. This condition is called *resonance*. It occurs at only one frequency for a given coil and capacitor. That frequency is called the *resonant frequency*. The power factor of a circuit operating at resonance is 1.

Practice Exercise.   Calculate the total reactance and phase angle of a 22-μF capacitor and a 319-mH coil connected in series with a 60-Hz source.

COMPONENTS USED WITH ELECTROMECHANICAL DEVICES—THE REAL WORLD

All too often a student is left thinking of resistors as zigzag lines, of coils as curlicues, and so on. This can cause problems when the time comes to work on real equipment. So now we will see what these devices are really like.

Resistors

Resistors come in several different styles. The most common ones are made of a carbon-composition material encased with ceramic. They are available in power ratings of $1/4$, $1/2$, 1, and 2 W. Figure 1-13 is an actual size-size photograph of several carbon-composition resistors. The value of these resistors is indicated by a series of colored bands, as shown in Fig. 1-14. In addition to the rated resistance, the code indicates the *tolerance*. The tolerance of a component is the maximum amount by which the actual value can be different from the specified value; a 100-Ω 10% resistor can have an actual value ranging from 90 to 110 Ω.

The second-most-common construction method for resistors is wire-wound. These are usually used in applications requiring a lot of power-handling capability. Their value and power rating may be color-coded or may be stamped onto the resistor body. Resistors for precision applications are often made of carbon film or metal film evaporated onto a small ceramic cylinder

Figure 1-13 Carbon-composition resistors. (Courtesy of Allen-Bradley Co.)

¼-watt
½-watt
1-watt
2-watt

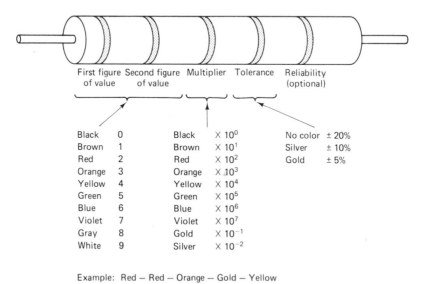

First figure of value   Second figure of value   Multiplier   Tolerance   Reliability (optional)

Black	0		Black	$\times 10^0$		No color	$\pm 20\%$
Brown	1		Brown	$\times 10^1$		Silver	$\pm 10\%$
Red	2		Red	$\times 10^2$		Gold	$\pm 5\%$
Orange	3		Orange	$\times 10^3$			
Yellow	4		Yellow	$\times 10^4$			
Green	5		Green	$\times 10^5$			
Blue	6		Blue	$\times 10^6$			
Violet	7		Violet	$\times 10^7$			
Gray	8		Gold	$\times 10^{-1}$			
White	9		Silver	$\times 10^{-2}$			

Example: Red — Red — Orange — Gold — Yellow
                      2   2 $\times 10^3$ $\pm 5\%$ 0.001% Failure rate

Figure 1-14 Resistor color code.

and encased in ceramic. For selecting a replacement resistor, then, four characteristics must be considered:

1. What value? (ohms)
2. What power rating? (watts)
3. What tolerance? (percent)
4. What construction? (carbon, wirewound, film)

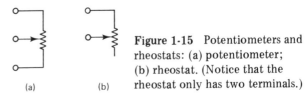

Figure 1-15 Potentiometers and rheostats: (a) potentiometer; (b) rheostat. (Notice that the rheostat only has two terminals.)

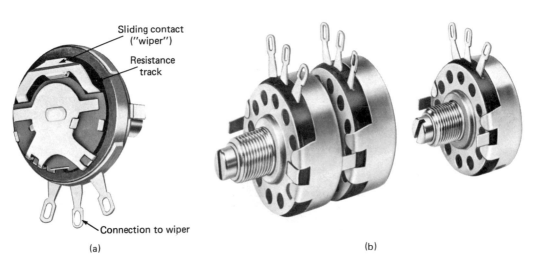

Figure 1-16 Potentiometers: (a) internal construction; (b) external view of single and dual units. (Courtesy of Allen-Bradley Co.)

Often it is necessary to vary the value of resistance in a circuit. This can be done by means of a *variable resistor*—either a *potentiometer* ("pot") or a *rheostat*. These are made of a "track" of resistive material, such as carbon, wire, or conductive plastic, contacted by a sliding "wiper." The track may be circular or linear. Potentiometers and rheostats are available in many values and in power ratings from $1/2$ W up to hundreds of watts. Figure 1-15 shows schematic symbols for a potentiometer (a) and a rheostat (b). Figure 1-16 shows the inside and outside of some common types of pots. To select a replacement pot or rheostat, you need to know:

1. What value?
2. What power rating?
3. What construction? (carbon, wirewound, etc.)
4. What shape? (linear "slide pot," rotary)

The value and power rating must usually be determined from the diagrams of the equipment being repaired; most pots and rheostats are not labeled.

Capacitors

There are two large families of capacitors: *electrolytic capacitors* and *electrostatic* (i.e., nonelectrolytic) *capacitors*. We'll talk about the electrostatic kind first. Almost all capacitors with values under 1 μF are made of two conducting plates separated by a dielectric of paper, ceramic, or plastic. Paper-dielectric capacitors are now obsolete except for one very important variety, the *oil-bath* or *paper-oil capacitors*. In them, the paper dielectric is impregnated with oil. The oil helps conduct heat to the outside of the capacitor case, where it can be dissipated; therefore, paper-oil capacitors are very useful for high-current ac use.

"Ah," you ask, "but why must capacitors get rid of heat if they do not dissipate power?" An astute question. Ideal capacitors dissipate no power. Real capacitors, however, dissipate a small amount of power because of several imperfections that are lumped together under the name "equivalent series resistance," or ESR. Oil-bath capacitors have a low ESR, but it is not zero. When current passes through them, some power is dissipated in the form of heat.

Ceramic capacitors are small, usually disk-shaped, and inexpensive. Plastic-dielectric capacitors may use either polyester or polystyrene as the dielectric, and are reasonably inexpensive and quite reliable. Figure 1-17 shows photographs of several paper-oil, ceramic, and plastic-dielectric capacitors.

Electrolytic capacitors use a chemical solution called an *electrolyte* to increase the capacitance available in a given physical size. This electrolyte works properly only if a difference in potential of the correct polarity is applied. Consequently, electrolytic capacitors are said to be *polarized*. All of this means that they have a + end and a − end, and must be connected correctly in order to work properly. In circuits that have no dc polarizing voltage available, *nonpolar* electrolytics may be used. These are really just two ordinary electrolytics connected "back-to-back" and enclosed in a single case. Photos of several polarized electrolytic capacitors are shown in Fig. 1-18a. Notice the + marking on the polarized capacitors. These indicate the proper polarity for connection in a circuit. Nonpolar electrolytics for high-current applications are also available. These are called ac electrolytics or

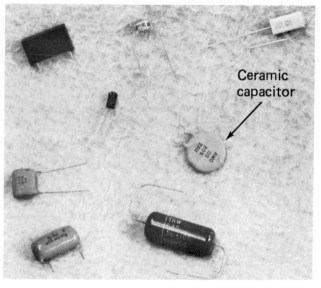

Ceramic
capacitor

(a)

(b)

Paper-oil

  Common values 0.1 to 10 μF

  High-current ac

  Reliable

  Expensive

Ceramic

  Common values 1 pF to 0.1 μF

  Low-current ac (control and
    amplifier circuits, etc.)

  Fairly reliable

  Inexpensive

Plastic-dielectric

  Common values 00.0005 to 4 μF

  Low-current ac

  Reliable

  Fairly inexpensive

Figure 1-17 Comparison of electrostatic capacitor types: (a) paper-oil
(courtesy of Aerovox, Inc.); (b) ceramic and plastic-dielectric (courtesy
of Sprague Electric Co. and TRW Capacitor Div.).

motor capacitors, and are constructed in the same way as regular nonpolars,
except that they are larger, to allow better heat dissipation. An example of a
motor-starting capacitor is shown in Fig. 1-18b. For continuous high-current
use, though, oil-bath capacitors are preferred because they have a lower ESR;
hence there is less heat to dissipate. Ac electrolytics, being less expensive, are
used for intermittent high-current applications. Electrolytics are available in
values from about 0.5 μF to over 1 F. They are moderately reliable, but do
deteriorate with age. Standard electrolytics are reasonably inexpensive. However, there is one type of electrolytic that is more stable, more reliable, and
has lower dc leakage through the dielectric. Naturally, these *tantalum* electrolytics are more expensive.

All capacitors have one very important rating besides the capacitance.
This is the *voltage rating*, which is the minimum *breakdown voltage of the*

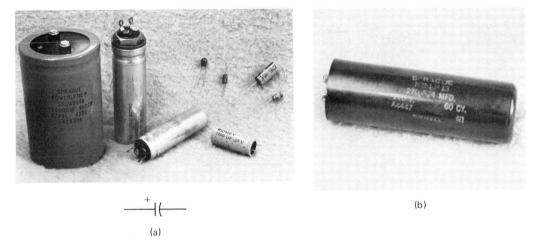

Figure 1-18 Electrolytic capacitors: (a) polarized electrolytics; (b) nonpolar electrolytic. (Courtesy of AT&T Technologies, Inc., Richey Electronics, and Sprague Electric Co.)

dielectric. This is often called the "working voltage," or WVDC. When a capacitor is replaced, then:

1. The capacitance (μF) must match.
2. The voltage rating of the new unit must be equal to or greater than that of the old one.
3. The type of construction should be the same as for the old unit. (*Exception*: For frequencies below 10 KHz plastic-dielectric capacitors can generally be used to replace any other type except paper-oil.)
4. The new capacitor *must* be installed with the correct polarity.

Failure to observe points 2 and 4 are likely to result in the explosion of the new capacitor, a spectacle that may prove amusing to coworkers, but can be dangerous and is *not* guaranteed to amuse supervisors.

Switches

Switches make and break circuits; that is, they turn things on and off. They are identified by how many circuits they turn on or off at a time (number of *poles*) and by how many "on" positions they have (number of *throws* or *positions*). Thus a double-pole, double-throw (DPDT) switch can simultaneously connect each of two circuits in either of two ways. Figure 1-19 shows photos and schematic symbols for a number of different varieties of switches.

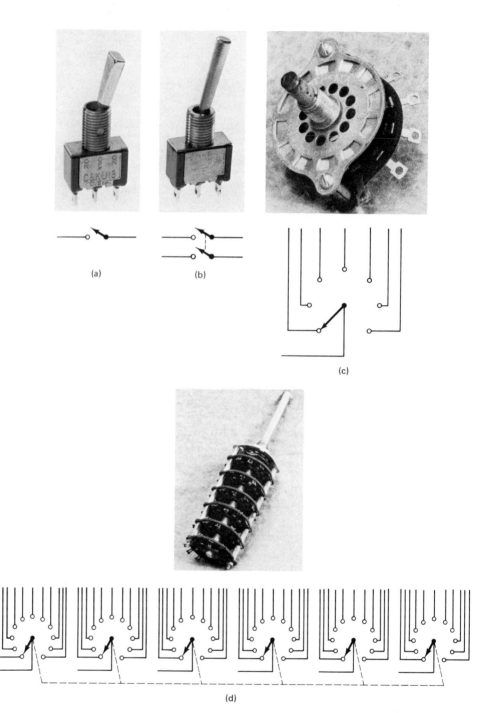

(a)

(b)

(c)

(d)

Figure 1-19 Switches: (a) SPST; (b) DPDT (dashed line indicates that both contacts operate at the same time); (c) one-pole, seven-position rotary selector switch; (d) six-pole, eleven-position rotary selector switch. [Photo (a) courtesy of C&K Components, Inc.)]

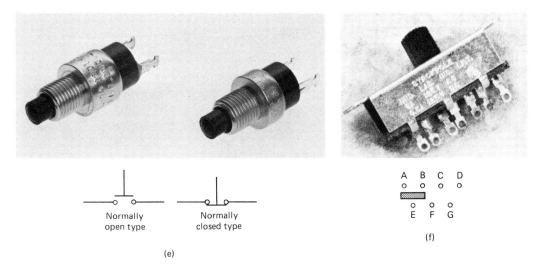

Normally
open type

Normally
closed type

A B   C   D

E   F   G

(e)

(f)

Figure 1-19 (cont.) Switches: (e) SPST pushbutton switches; (f) slide-type selector switch (connects together three terminals at a time). [Photos (e) courtesy of C&K Components, Inc.; (f) courtesy of Stackpole Components Div.]

Looking at this figure, you will discover another way to define poles and throws. The number of poles of a switch is equal to the number of moving contacts. The number of throws is the number of fixed contacts per moving contact. Thus a switch with 10 fixed contacts and two moving contacts would be a two-pole, five-throw (or two-pole, five-position) switch. Even though switches are fairly simple to understand, there are so many varieties, and switches are so common, that it is worth spending some time with Fig. 1-19 to make sure that you understand it clearly.

Switches are rated not only by their configuration (poles and throws), but also by the maximum current and voltage that the contacts can handle. Too high a voltage may arc across the contacts when they are supposed to be open. Too high a current may overheat the contacts and burn (oxidize) them or even weld them together.

Inductors

Inductors are used in all areas of electronics, and therefore come in many shapes and sizes. However, those used in connection with electromechanical devices are primarily power-supply filter chokes or radio-frequency (RF) chokes. Filter chokes make use of inductors' ability to pass dc while presenting a reactance to ac. Thus in a dc power supply any ac that happens to be present along with the dc is blocked from reaching the load. Filter chokes

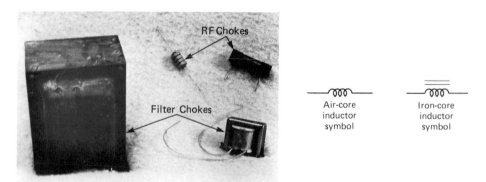

Figure 1-20 Inductors.

generally look like the ones in Fig. 1-20. They are rated in terms of inductance and current. RF chokes are used in circuits that generate radio interference to prevent that interference from being fed onto the power line.

Transformers

Transformers have not really even been introduced yet, so we had better do
that first. If two separate coils of wire are placed near each other or wound
on the same core, the magnetic field of each will affect the other. (This phenomenon is called *mutual induction.*) What's more, if coil A has 100 turns
and coil B has 200 turns, and 5 V ac is applied to coil A, coil B will have
10 V at its terminals. Then if a 1-A load is connected to coil B, 2 A will be
drawn from the source that feeds coil A. In other words, a transformer can
change ac voltage and current levels; *if it steps up voltage, it steps down current, and vice versa.* Figure 1-21 illustrates transformer operation. In any
transformer

$$\frac{V_P}{V_S} = \frac{N_P}{N_S} = \frac{I_S}{I_P}$$

where V_P = primary voltage

V_S = secondary voltage

I_P = primary current

I_S = secondary current

N_P = number of turns in the primary winding

N_S = number of turns in the secondary winding

Notice that it is not necessary to know the number of turns; if the voltages
are known, the currents can be calculated, and vice versa.

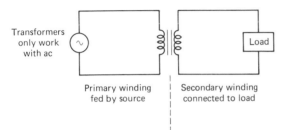

Transformers only work with ac

Primary winding fed by source

Secondary winding connected to load

Load

Figure 1-21   Transformer operation.

Transformers are rated for operation at certain frequencies; the ones used in power circuits for motors, generators, and relays are usually rated for either 50 or 60 Hz, although 400-Hz units are available for some aircraft applications. Also, they are rated in terms of voltage output and current capability. Power transformers normally have 117- or 230-V primaries, and the rated secondary voltages are based on the assumption that the primary is being fed the rated voltage. The maximum secondary current is also specified.

A peculiar transformer-like creature is the *autotransformer*. These are really just tapped inductors, but they can be used like transformers in stepping voltage up or down. Many autotransformers are variable. Figure 1-22 shows a photograph of some common transformers and a variable autotransformer (the unit at the far left).

Figure 1-22   Typical transformers.

Rectifiers

Virtually all homes and factories are supplied with ac power. Yet much equipment requires dc. The devices that change ac into dc are called rectifiers. The way a rectifier changes ac to dc is that it passes electrons in only one direction. This concept is shown in Fig. 1-23.

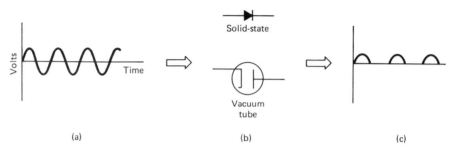

Figure 1-23   Rectification: (a) input waveform; (b) schematic symbol; (c) output waveform.

There are a number of different types of rectifiers. In most recently built equipment, solid-state rectifiers are used. Older equipment often used selenium rectifiers or vacuum-tube rectifiers. Very high current equipment may use gas-filled tube rectifiers. Photographs of these various devices are shown in Fig. 1-24. Just remember, though, as different as they may look, all

Figure 1-24   Rectifiers: (a) high-current mercury-vapor rectifier tube; (b) (top to bottom) solid-state diode, bridge rectifier, full-wave rectifier; (c) vacuum-tube rectifiers.

rectifiers do the same job—converting ac to dc. Rectifiers have two important ratings: maximum reverse voltage and maximum forward current. Any rectifier can be replaced by one having the same or higher voltage and current ratings. These ratings are seldom printed on the rectifier. Instead, a manufacturer's type number is stamped on the case. This number can be looked up in an electronic parts distributor's catalog or cross-referenced in one of the replacement semiconductor guides that are published by a number of manufacturers. Often four rectifiers are enclosed in a single case. These are called *bridge rectifiers*.

A device very similar to a rectifier is the *zener diode*. In appearance, a zener diode is not distinguishable from a rectifier. At low voltages, the performance is no different either. But at some specific voltage, the zener diode will begin to pass current in the "reverse" direction. This is called the diode's *zener voltage*. The well-defined reverse breakdown characteristic makes zener diodes useful for circuits that regulate voltage. Zener diodes are rated in terms of zener voltage and power dissipation. A defective zener diode can be replaced by a unit having the same zener voltage and equal or greater power ratings.

Transistors and Related Devices

The operation of transistors, silicon-controlled rectifiers (SCRs), and triacs is introduced in Chapter 3. At this point, though, we need to discuss briefly the ratings of these devices and show photos of what they look like (Fig. 1-25). All three of these—transistors, SCRs, and triacs—have a maximum (breakdown) voltage and operating current. Transistors have, in addition, a gain and a maximum power dissipation. A replacement device should have equal or greater ratings in all of these categories except gain, which should be

Figure 1-25   Solid-state control devices.

matched as closely as possible. Like rectifiers and zener diodes, transistors and their SCR and triac cousins have manufacturer's type numbers stamped on their cases. These numbers must be looked up in a catalog or cross-referenced in a replacement guide if the ratings are to be found.

Fuses and Circuit Breakers

Very often, a defect in one component in a circuit can cause other components to fail. For example, a shorted transistor or diode can draw too much current and burn out a transformer. To prevent this, there are two types of "overcurrent protection devices." The most common is the plain ol' fuse (see Fig. 1-26). Fuses are available in many shapes and in virtually any current rating. Most fuses blow very rapidly if too much current is drawn. There is a type of fuse, however, that will not blow unless the high current continues for a certain period of time. These *slow-blow* fuses are used with equipment that may draw very brief surges of high current. Fuses are also rated in terms of voltage. This rating specifies the maximum voltage that can be applied without jumping the gap that results when the fuse blows. Thus using a 32-V

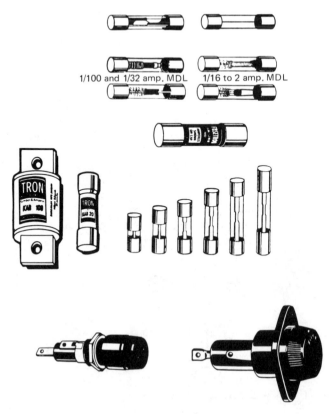

Figure 1-26 Fuses. (Courtesy of Bussmann Div., McGraw-Edison Co.)

fuse in a 500-V circuit is fine—until the fuse blows. Then the fuse may not do its job properly, because the gap that melted in the fuse wire may be narrow enough for 500 V to arc across.

The other overcurrent protection device is the circuit breaker. Circuit breakers are magnetically or thermally operated devices that do the same job that fuses do; that is, they open a circuit when too much current is drawn. Unlike a fuse, though, a circuit breaker can be reset. Normally, their resetability makes circuit breakers less expensive than fuses in the long run, although they cost much more initially. Circuit breakers have the same ratings as fuses: maximum current and maximum voltage. Figure 1-27 shows a number of circuit breakers.

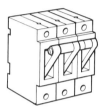 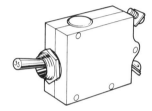

Figure 1-27   Circuit breakers.

BASIC DC POWER-SUPPLY CIRCUITS

Many motor control circuits and almost all relay circuits are fed by a circuit that converts the 117-, 220-, or 440-V ac supplied by the power company into dc of some other voltage. We have already discussed all the components that are used in such a dc power supply. Below the block diagram in Fig. 1-28 are shown actual circuit examples that indicate how the components could be connected in a real piece of equipment.

SUMMARY

1. A series resistive circuit has:
   a. No inductors
   b. No capacitors
   c. Only one current path
   d. The sum of all voltage drops equal to the sum of all applied voltages.
   e. A total resistance equal to the sum of all individual resistances
   f. The same current at every point in the circuit
   g. Current equal to $E_{\text{applied}}/R_{\text{total}}$
   h. A total power dissipation equal to the sum of the individual resistors' power dissipation

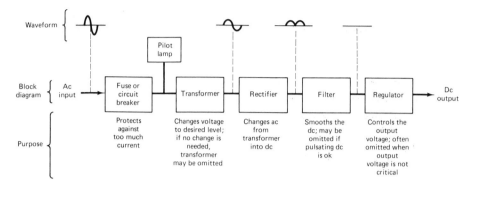

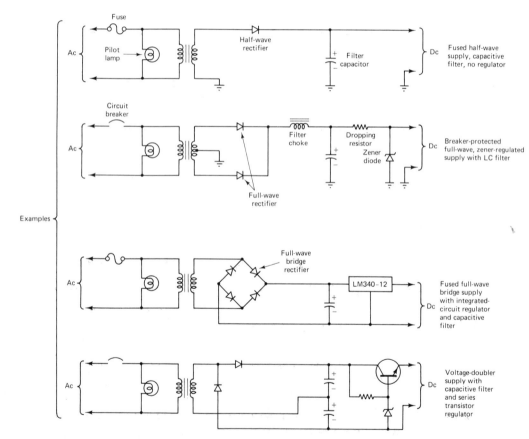

Note: These are only examples. Any combination of rectifier,
filter, and regulator may be used. Some regulator circuits
are complex. These are intended as a road map for
common types of power supplies.

Figure 1-28 Dc power supplies.

2. A circuit's conductors may all be wires, or a part of the circuit may be completed by the chassis of the equipment. The latter type of circuit is called a ground-return circuit.

3. Voltages in a circuit are usually measured with respect to the circuit's common or ground point; usually this is the chassis.

4. Voltages are measured by paralleling the voltmeter with the components whose voltage drop is to be found.

5. Current is measured by placing the ammeter in series with the circuit being tested.

6. Two or more resistors in series form a voltage divider. The voltage-division rule is extremely handy for calculating drops in a voltage-divider circuit.

7. Power dissipated by a resistor is the product of the voltage across that resistor and the current through it.

8. Voltage sources connected in series with the same polarity are series-aiding; sources connected in series with opposite polarities are series-opposing. The total voltage of a group of series sources is equal to the sum of the series-aiding sources minus the sum of the series-opposing sources.

9. Parallel circuits have:
   a. More than one current path
   b. The same voltage across each branch
   c. Branch currents given by $E_{applied}/R_{branch}$
   d. A total current equal to the sum of the branch currents
   e. A total resistance given by

   $$R = \cfrac{1}{\cfrac{1}{R_1} + \cfrac{1}{R_2} + \ldots + \cfrac{1}{R_n}}$$

   f. A total resistance lower than the lowest branch resistance
   g. A total power dissipation equal to the sum of the branch powers

10. The current capacity of equal voltage sources adds when the sources are connected in parallel. Connecting unequal voltage sources in parallel may damage the voltage sources.

11. Combination circuits have both series and parallel parts. Voltages and currents of a combination circuit can be calculated if the circuit is broken up into its series elements and its parallel elements; then each of these is reduced to a single equivalent resistance.

12. Inductors cause the voltage to lead the current by 90°. Capacitors cause the current to lead the voltage by 90°. Eli the Iceman can help you to keep this straight:

E	L		I	the	I	C		E	man
v	i		c		c	c		v	
o	n		u		u	a		o	
l	d		r		r	p		l	
t	u	in	r		r	a	in	t	
a	c	an	e		e	c	a	a	
g	t	comes before	n		n	i	comes before	g	
e	o		t		t	t		e	
	r					o			
						r			

13. Reactance is the ratio of voltage to current in a resistanceless ac circuit. Current in a resistor produces heat; current in a reactive component does not.

14. Inductive reactance is given by $|X_L| = 2\pi f L$.

15. Voltages, currents, resistances, and reactances in ac circuits are phasors, which means that they have both magnitude and phase angle. Inductive reactance has a phase angle of $90°$.

16. All reactances combine in series or parallel just as resistors do.

17. Capacitors cause the voltage to lag the current by $90°$. Capacitive reactance is given by

$$|X_C| = \frac{1}{2\pi f C}$$

Capacitive reactance has a phase angle of $-90°$.

18. Capacitances in parallel add. Capacitors in series combine in the way that resistors in parallel do.

19. The total reactance in a circuit is

$$|X_T| = |X_L| - |X_C|$$

and has the phase angle of whichever individual reactance is larger.

20. Reactive and resistive voltage drops add according to

$$|E_{total}| = \sqrt{|E_x|^2 + |E_R|^2}$$

21. Impedance is the ratio of voltage to current in an ac circuit that has both resistance and reactance. It is a phasor. Its magnitude is found from

$$|Z| = \sqrt{(|X_L| - |X_C|)^2 + R^2}$$

Its angle is found from

$$\theta_Z = \tan^{-1}\left(\frac{|X|}{R}\right)$$

TABLE 1-2 Characteristics of Components

Generic name	Ratings	Construction	Values	Notes
Resistors	Ohms, watts, tolerance (%)	Carbon composition	$1\ \Omega$ to $22\ M\Omega$	General purpose
		Wirewound	Up to $110\ k\Omega$	High power and precision
		Carbon-film	$10\ \Omega$ to $1\ M\Omega$	Precision
		Metal-film	$10\ \Omega$ to $1\ M\Omega$	Precision
Pots and rheostats	Ohms, watts	Carbon composition	$50\ \Omega$ to $500\ k\Omega$	General purpose
		Wirewound	$10\ \Omega$ to $50\ k\Omega$	High power
		Conductive plastic	$500\ \Omega$ to $5\ M\Omega$	Long life
Capacitors	μF or pF, volts (WV dc), tolerance	Paper-oil	3 to $60\ \mu F$	High ac current
		Ceramic	$10\ pF$ to $0.1\ \mu F$	Inexpensive, small
		Plastic	0.001 to $5\ \mu F$	High quality, general purpose
		Standard electrolytic	1 to $100,000\ \mu F$	General purpose
		Tantalum electrolytic	1 to $500\ \mu F$	Precision or high reliability
Switches	Number of poles, number of throws or positions, current, voltage	Toggle	—	At most three positions (throws)
		Slide	—	Up to five positions
		Pushbutton	—	Not over two positions
		Rotary	—	Virtually any number of positions
Inductors	Henries, current	—	—	—
Transformers	$V_{primary}$, $V_{secondary}$, and $I_{secondary}$ or volt-amperes, $V_{primary}$ and $V_{secondary}$	—	—	—
Rectifiers	Voltage	Single	1 A to 1 kA,	—
	Current	Bridge	50 V to 20 kV	Multiple rectifiers in a single case
Zener diode	Voltage		2 to 200 V	Voltage regulators
	Power		$\frac{1}{2}$ to 50 W	
Transistors	Voltage, current, power, gain		—	—
SCRs, triacs	Voltage, current		—	—
Fuses, circuit breakers	Current, voltage		—	Fuses may be either fast-blow or slow-blow

22. Apparent power is the product of the applied voltage and current in a circuit containing reactance.

23. True power is the actual power dissipated by resistive elements.

24. Power factor is the ratio of true power to apparent power; it is also given by PF $= \cos \theta_Z$.

25. When capacitive and inductive reactances are equal, a circuit is resonant.

26. Information on components is summarized in Table 1-2.

27. Dc power supplies convert ac at the input to dc at the output. Often, they change voltages in the process. A basic dc power supply includes a transformer to change ac voltages, a rectifier to change ac to dc, and a filter to remove the remaining ac from the dc. It may contain a voltage regulator to keep the output voltage constant.

QUESTIONS

1. In Fig. P1-1, what is the voltage drop across R_3?

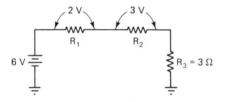

Figure P1-1

2. What is the total current in the circuit shown in Fig. P1-1? the total resistance? the resistance of each resistor?

3. What is the current in the 10-Ω resistor in the circuit shown in Fig. P1-3? the current in the 15-Ω resistor?

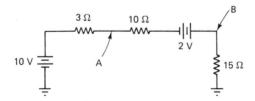

Figure P1-3

4. What is the voltage at point A in the circuit of Fig. P1-3? at point B? What is the voltage drop across the 15-Ω resistor?

5. How would an ammeter be connected to measure the current in the 3-Ω resistor in the circuit of Fig. P1-3?

6. What are the voltages at A, B, and C in the circuit shown in Fig. P1-6? (*Hint:* Use the voltage-division rule.)

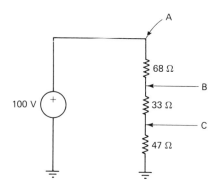

Figure P1-6

7. How much power is dissipated by each resistor in the circuit of Fig. P1-6?

8. What is the current in the resistor in the circuit shown in Fig. P1-8?

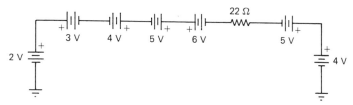

Figure P1-8

9. What is the total current in the circuit shown in Fig. P1-9?

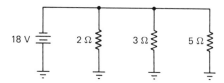

Figure P1-9

10. What is the voltage across the 2-Ω resistor of Fig. P1-9? across the 5-Ω resistor? What is the total equivalent resistance?

11. What is the total power dissipated by the circuit of Fig. P1-9?

12. Which of the following answers is the total resistance of the circuit shown in Fig. P1-12?   (1) 17 Ω; (2) 20 Ω; (3) 1.5 Ω; (4) 2 Ω; (5) 1 Ω.

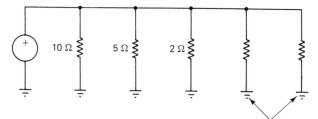

Markings on these two resistors were too scorched to read; you will have to solve the problem without knowing their values   **Figure P1-12**

13. Why are batteries often connected in parallel?

14. Find the total current in each circuit of Fig. P1-14.

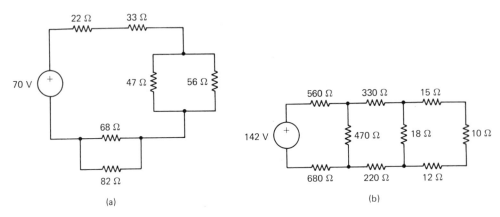

Figure P1-14

15. What is the power dissipated in the 470-Ω resistor in circuit (b) of Fig. P1-14?

16. Are the voltage drops shown in the circuit of Fig. P1-16 possible? Why or why not?

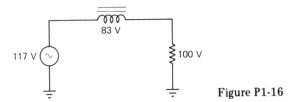

Figure P1-16

17. If the inductor in the circuit of Fig. P1-16 is 2 H and the resistor is 100 Ω, what is the true power dissipated in each component? What is the power factor?

18. What would the reactance of a 2-H inductor be at a frequency of 6.6 Hz? at 60 Hz?

19. What is the impedance of the circuit shown in Fig. P1-19? (Remember, Z is a phasor.)

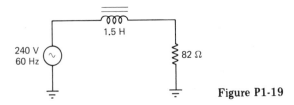

Figure P1-19

20. In Fig. P1-19, what would the impedance of the circuit be if the frequency were changed to 400 Hz?

21. What is the total inductance of the circuit shown in Fig. P1-21?

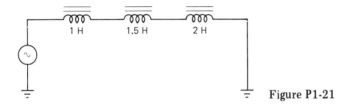

Figure P1-21

22. What is the total inductive reactance of the circuit shown in Fig. P1-22?

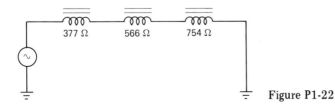

Figure P1-22

23. What is the total capacitance of each circuit in Fig. P1-23?

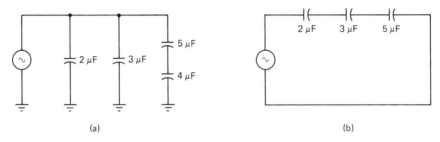

(a)   (b)

Figure P1-23

24. What is the capacitive reactance of each circuit in Fig. P1-24?

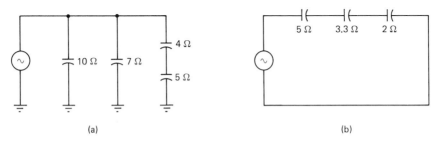

(a)   (b)

Figure P1-24

25. What is the reactance of the circuit shown in Fig. P1-25? the impedance?

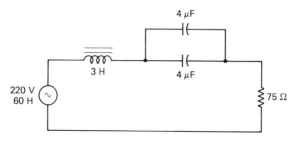

Figure P1-25

26. In Fig. P1-25, what would the impedance of the circuit be at a frequency of 32.5 Hz? What is this condition called?

27. Would it be okay to replace:
    (a) A wirewound resistor with a carbon-composition resistor?
    (b) A carbon-composition resistor with a wirewound resistor?
    (c) A plastic capacitor with a ceramic capacitor?
    (d) A standard electrolytic capacitor with a tantalum capacitor?
    (e) A paper-oil capacitor with a standard electrolytic capacitor?

28. A replacement resistor should have the same _____ and equal or higher _____.

29. A replacement capacitor should have the same _____ and equal or higher _____.

30. A replacement transformer should have the same _____ and _____ and equal or higher _____.

31. A replacement inductor should have the same _____ and equal or higher _____.

32. A replacement rectifier should have equal or greater _____ and _____ ratings.

33. What characteristics of a transistor must be matched or exceeded by a replacement unit if an exact replacement cannot be obtained?

34. A replacement SCR or triac should have equal or greater _____ and _____ ratings.

35. Why is it not acceptable to replace a fuse with a "heavier-duty" one, to keep it from blowing?

36. What are the sections of a dc power supply, and what is the purpose of each section?

2

MAGNETISM

If you are a science-fiction fan, chances are that you have read at least a few stories in which an "invisible force field" played an important part. These "fields" have been a favorite gimmick of science-fiction writers for decades. Scientists today actually know of at least three types of invisible force fields. The one you are most familiar with is the gravitational field. The second is the one that causes electrons to move along a wire, and causes cellophane to be hard to let go of — the electric field. The third is that "mysterious phenomenon" known as the magnetic field.

Before we begin discussing the magnetic field, I should point out that the purpose of this chapter is to give you the background you need to understand how electromagnetic devices operate. In doing so, a number of equations will be used. The important thing for you to learn from the equations is the *relations* expressed between the variables. It is unlikely that you will ever need to do any calculations with these equations. Also, several terms will be introduced. My purpose in doing so is to give you the vocabulary to understand discussions later in the book. In other words, read this chapter for a general understanding, not for detail.

Well then, what really is a field? *A field is a region in which a force is exerted without physical contact.* If this field affects only certain kinds of metals, and it affects wires that are carrying electric current, it is a magnetic field. To make it easier to visualize the magnetic field, a concept called "lines of force" was invented. To represent a strong field, lines of force are shown close together. For a weak field, they are shown spaced far apart. Lines of force, or *flux*, are represented by the symbol ϕ (Greek lowercase letter phi)

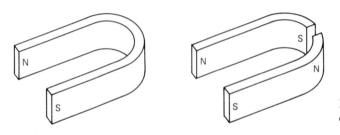

Figure 2-1 North and south poles
of a magnet.

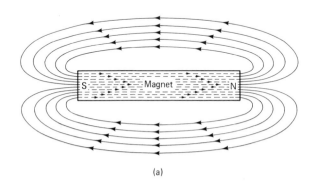

(a)

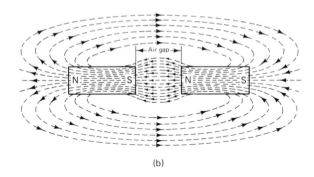

(b)

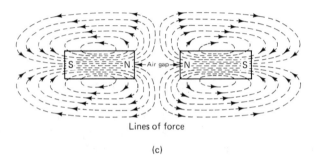

Lines of force

(c)

Figure 2-2 Magnetic fields of
permanent magnets: (a) bar
magnet; (b) unlike poles attract;
(c) like poles repel.

and are measured in *webers* (meter-kilogram-second system), or *maxwells* (centimeter-gram-second system). It is important to remember that lines of force are purely imaginary; we speak as if they were real to aid understanding, but they do not actually exist.

As every schoolchild knows, every magnet has two locations where the field is strongest. These are called the *poles*, and each is named *north pole* or *south pole*, according to which end is attracted by the earth's magnetic north pole. A fundamental law concerning magnetic poles is that *like poles repel and unlike poles attract each other.* (This brings up the point that the earth's magnetic north pole is actually a south pole, but this fact is usually of value only for amazing friends at parties.) If we cut a magnet in half, we cannot separate it into a "north half" and a "south half." What happens instead is that two new poles are created. This situation is diagrammed in Fig. 2-1.

The use of lines of force to show the concentration of a magnetic field at the poles is shown in Fig. 2-2. Since the magnetic field is a vector quantity, the lines of force are usually drawn with arrowheads showing their direction. The magnetic field's direction is from north pole to south pole. Notice that the lines of force are closer together at the poles, and farther apart everywhere else. *The number of lines of force that pass through a given cross-sectional area is referred to as the flux density*, identified by the symbol B:

$$B = \frac{\phi}{A} = \frac{\text{number of lines of force}}{\text{unit area}}$$

As noted above, ϕ is measured in maxwells or webers. Thus the centimeter-gram-second unit of flux density is the *gauss*: one maxwell per square centimeter = 1 *gauss*. The meter-kilogram-second unit of flux density is the *tesla*: one weber per square meter = 1 *tesla*. Conversely, total flux is given by

$$\phi = B \times A$$

ELECTROMAGNETISM

Around 1819, Hans Christian Oersted discovered that whenever an electric current passes through a conductor, a magnetic field results. We now know that the conductor is not necessary. Even if a stream of electrons is moving through a vacuum, a magnetic field will still be set up. Thus we can describe a law of electromagnetism: *Whenever an electric charge is in motion, a magnetic field will result.* The direction of the magnetic field is given by the left-hand rule: If you point the thumb of your left hand in the direction of electron motion, the fingers will indicate the direction of the magnetic lines of force. Those lines of force are concentric with the path of the charge (see Fig. 2-3). If the current path is a loop rather than a straight line, the resulting field will be as shown in Fig. 2-4. In part (a) of the figure, the curved lines represent lines of force near the wire. These lines add together in the center

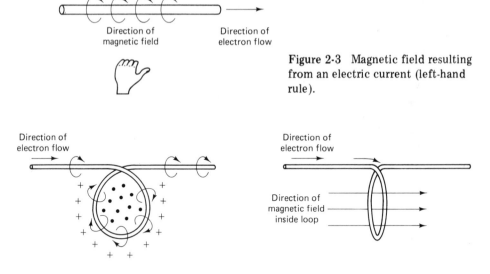

Figure 2-3   Magnetic field resulting from an electric current (left-hand rule).

Figure 2-4   Magnetic field around looped conductor: (a) side view; (b) edge view.

of the loop to produce a concentrated field pointing up out of the page (represented by dots). Outside the loop, the lines of force point into the page ("tailfeathers of arrows," shown as + +). If we then group a number of loops together to form a helical (corkscrew-shaped) coil, the field will add on the inside and outside of the coil to produce an electrical equivalent of a bar magnet, as shown in Fig. 2-5. An electromagnet of this kind is called a *solenoid*. The left-hand rule can be used for coils also. If the fingers of the left hand point in the direction in which a coil is wound (i.e., the current path), the thumb will point in the direction of the magnetic field.

The magnetic effect of the electric current in a coil is called the *magnetomotive force*, or simply the *mmf*, and is represented by the symbol $\mathfrak{F}$. Mmf depends on the number of turns in a coil and the current in the coil:

$$\mathfrak{F} = NI$$

Mmf is measured in ampere-turns.

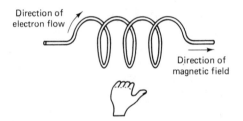

Figure 2-5   Magnetic field around coil (left-hand rule).

FERROMAGNETISM

So far, we have considered magnets that either were in a universe by themselves or were in the presence of other magnets. Now let's see what happens to the magnetic field when another material is immersed in it. Obviously, there are three possible effects:

1. The material can weaken the field.
2. The material can strengthen the field.
3. The material can leave the field unaffected.

Figure 2-6 illustrates these effects. Only a vacuum (i.e., no material at all) will leave the field unaffected. Materials that weaken the field are called *diamagnetic*. Materials that strengthen the field fall into two classes. Those that strengthen the field only very slightly are called *paramagnetic*. Those that strengthen the field significantly are called *ferromagnetic*. The paramagnetic and diamagnetic effects are so slight that these two classes of material are often referred to as nonmagnetic. They comprise most kinds of matter. Ferromagnetic materials include primarily iron, nickel, cobalt, and their alloys.

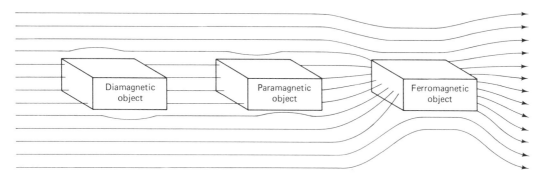

Figure 2-6 Effect of material objects on magnetic field intensity. (Effects of diamagnetism and paramagnetism are exaggerated for clarity.)

The magnitude of a material's effect on a magnetic field is described by its *permeability* (symbolized by the Greek lowercase letter mu, μ). Permeability can be considered to be the factor by which the flux density in a field is multiplied when the ferromagnetic material is introduced. This is illustrated in Fig. 2-7. Vacuum has no effect on a magnetic field, so its permeability is 1. Diamagnetic materials have permeabilities less than 1. Paramagnetic materials have permeabilities greater than 1; and ferromagnetic materials, much greater than 1. Some examples of permeabilities are given in Table 2-1.

The permeability of a given ferromagnetic material is not a fixed num-

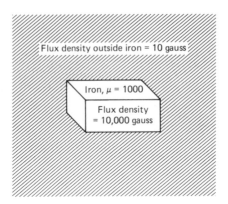

Figure 2-7 Concentrating effect of ferromagnetic material.

TABLE 2-1 Permeabilities of Some Common Materials

Material	Permeability	
Copper	0.999	(diamagnetic)
Air	1.0000004	(paramagnetic)
Iron	200	
Nickel	100	(ferromagnetic
Mu metal (alloy of nickel, chromium, copper, and iron)	20,000	permeability varies with magnetic field strength)

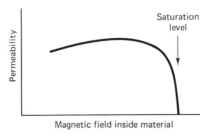

Figure 2-8 Variation of permeability with magnetic field strength.

ber. It depends on the strength of the field to which the material is exposed. The variation is illustrated in Fig. 2-8.

Another way of looking at the effect of permeability is to draw a graph of the strength of the magnetic field inside a material, plotted against the strength of the field surrounding the material. This graph shows a peculiar, fat-S shape and is referred to as a *hysteresis* curve. A typical hysteresis curve is shown in Fig. 2-9. The curve can be explained as follows:

1. The iron is initially unmagnetized. As the magnetic field is applied and increased, the field inside the iron increases to its maximum possible level (saturation).

2. The iron holds some magnetism as the field is reduced. With zero field applied, some magnetism remains.

3. The applied field's polarity is reversed, but some reverse field is needed to bring the field inside the iron down to zero. As the strength of the reversed field increases, the iron becomes magnetized in the reverse direction. The field inside the iron increases until saturation occurs with the reversed polarity.

4. As the applied field strength is reduced, the field inside the iron decreases, but the iron still has some internal field even when the applied field is zero.

5. If the external field is now applied again with the original polarity, then increased, the field in the iron will also be returned to its original polarity and increased until saturation occurs.

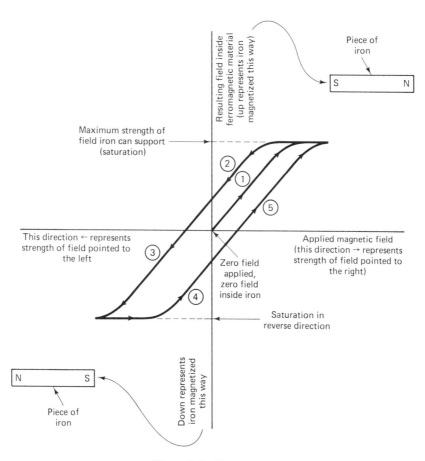

Figure 2-9   Hysteresis.

For a certain piece of material, the permeability and the size and shape of the material combine to affect a magnetic field by a specific amount. This amount is stated as the *permeance* of the material. Permeance, symbolized by $\mathcal{P}$, is given by

$$\mathcal{P} = \frac{\mu A}{l}$$

where A = cross-sectional area of the material

l = length of the material

μ = permeability

The inverse of permeance is called *reluctance*, $\mathcal{R}$:

$$\mathcal{R} = \frac{1}{\mathcal{P}} = \frac{l}{\mu A}$$

Reluctance can be thought of as corresponding to the difficulty of magnetizing a given piece of material. Permeance is measured in henries; reluctance, in (henries)$^{-1}$.

The magnetomotive force, total flux, and reluctance are related to each other by an equation that is often called the "magnetic Ohm's law":

$$\mathcal{F} = \phi\mathcal{R}$$

From it we derive the equivalent equations $\mathcal{R} = \mathcal{F}/\phi$ and $\phi = \mathcal{F}/\mathcal{R}$. In this equation, we see that $\mathcal{F}$ can be compared to voltage in an electric circuit; ϕ, to current; and $\mathcal{R}$, to resistance. This equation is important for an understanding of magnetic concepts, even though its use by technicians for calculation is nil. As an example of the application of the *concept* of the magnetic Ohm's law, let's consider a loudspeaker's magnetic structure, as shown in Fig. 2-10. Notice that there is an almost complete path for the flux (lines of

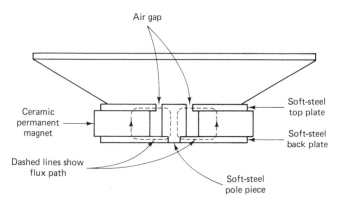

Figure 2-10 Magnetic circuit of loudspeaker.

force) through the magnetic material. This path is called the *magnetic circuit*. Ideally, the only significant reluctance in the circuit is the air gap in which the voice coil travels. (Actually, there are microscopic gaps between the top plate and the magnet, between the magnet and back plate, and between the back plate and pole piece, but usually these are ignored in practice.) The reluctance of magnetic materials is very low. Thus if a smaller air gap is used (less reluctance), all other things being equal, the flux increases:

$$\phi = \frac{\mathcal{F}}{\mathcal{R}}$$

$\mathcal{F}$ ←—If this stays the same,

$\mathcal{R}$ ←—and this gets smaller,

—then this must increase.

This principle is also important in the magnetic circuits used in relays, motors, and other devices.

An important characteristic of a ferromagnetic material is its *retentivity* or ability to retain magnetism. Another word that is sometimes used for retentivity is *remanence*. Both words mean the same thing. A material with a high retentivity tends to hold its magnetism quite well. Such a material is sometimes called a "hard" magnetic material, and would be used for making permanent magnets. Materials having low retentivities lose their magnetism rapidly when removed from the presence of a magnetic field. Such materials are said to be magnetically "soft," and are used in applications in which permanent magnetism is not wanted. Retentivity can also be understood as a kind of memory of previous fields to which a magnetic material has been exposed. The retentivity of a material shows up as the "fatness" of its hysteresis loop, as in Fig. 2-11.

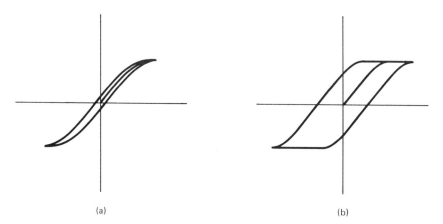

(a)   (b)

Figure 2-11 Hysteresis loops for two materials of different retentivity: (a) hysteresis loop of soft iron that has a poor "memory"—and therefore would make a poor permanent magnet; (b) hysteresis loop of hard steel that has a good "memory"—and therefore would make a good permanent magnet.

The magnets in loudspeakers and other devices that use permanent magnets are usually supplied to the manufacturer unmagnetized. They are then magnetized by means of a powerful electromagnet. Barium ferrite and other ceramic magnets usually hold this magnetism unless they are deliberately demagnetized by application of a strong reverse field. Alnico and tempered steel are somewhat subject to accidental demagnetization, from any of three causes. First, for every magnetic material there is a temperature at which it will lose its magnetism. This is called the Curie temperature (named after Pierre Curie). For iron, this is 770°C; for nickel, it is 350°C. Permanent magnets that are heated to this temperature will become demagnetized. Second, metallic permanent magnets can be partially demagnetized as a result of mechanical shock (a 6-ft drop onto a concrete floor!). Third, under some conditions of use, the normal electromagnetic fields generated in the devices themselves can become large enough to partially demagnetize a magnet.

MAGNETIC FORCE ON A MOVING CHARGE

If an electron is moving in a stationary magnetic field, the electromagnetic field of the electron current will interact with the stationary field to produce a force on the electron. This situation is diagrammed in Fig. 2-12. In this figure, the large dots ("arrow points") represent a magnetic field pointing up out of the page. The small dots and + signs show the field resulting from the moving electron. Above the electron, the dots and +'s partially cancel, weakening the field. Below the electron the "arrow points" add, strengthening the field. The resulting force pushes the electron; *the direction of the force is always from the stronger to the weaker field.* If that electron is one of many that constitute an electric current in a wire, a force will be exerted on the wire. This is often called the *motor effect.* There are several practical examples. The first, of course, is the electric motor, for this is the principle on which a motor works. Other examples would be voltmeter movements, loudspeakers, or any other electromagnetic-mechanical device. A final example would be a TV picture tube, in which the electron beam that "paints" the picture is swept repeatedly from one side of the screen to the other, and from top to bottom, by the magnetic field of the deflection yoke.

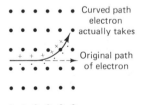

Figure 2-12 Magnetic force on a moving charge.

VOLTAGE INDUCED BY MAGNETISM

The electromagnetic effect is somewhat reversible. Just as an electric current and a stationary magnetic field can combine to produce a force that can cause motion, a magnetic field that is in motion with respect to a conductor will cause a voltage to be generated within that conductor (Fig. 2-13). This voltage is sometimes called an electromotive force or emf, even though it is not a true force. The faster the relative motion between the field and the conductor, the greater the voltage. The more of the conductor that is immersed in the magnetic field, the greater the voltage. These relations are expressed by

$$E_{induced} = B \times l \times v$$

where B = magnetic flux density

l = length of conductor

v = velocity of motion

Notice that it does not matter whether the conductor moves or the field moves; what matters is that there be relative motion between the two. Another way of expressing this same relationship is Faraday's law:

$$E_{induced} = \frac{-N d\phi}{dt}$$

where N is the number of turns in a coiled conductor and $d\phi/dt$ expresses the number of lines of force that the conductor cuts across each second. This effect is sometimes called the *generator effect*.

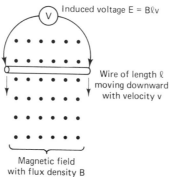

Induced voltage E = Bℓv

Wire of length ℓ
moving downward
with velocity v

Magnetic field
with flux density B

Figure 2-13 Magnetically induced voltage.

COUNTERVOLTAGE

Motor effect and generator effect can occur at the same time. When a current-carrying conductor is in a magnetic field, the resulting magnetic force will cause the conductor to move. Motion of the conductor in the magnetic

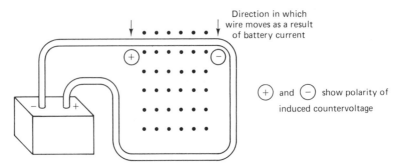

Figure 2-14   Countervoltage.

field will result in the generation of a voltage in the conductor. *The polarity of the voltage will be opposite that of the voltage that resulted in the original motion* (Fig. 2-14). The resulting net voltage applied to the conductor is equal to the battery voltage minus the induced *countervoltage.* This countervoltage is sometimes called counter emf, and occurs in every motor, loudspeaker, or other device that converts electrical energy via magnetism to mechanical energy. It is countervoltage that is responsible for the phenomenon of inductance. The reason that the countervoltage *opposes* (polaritywise) the original voltage can be easily understood. Many people have dreamed of a device that would generate its own energy to run itself and also do useful work. Such a device could be called a *perpetual-motion* machine. However, in every mechanical device, some mechanical energy is converted to heat energy by friction. In every electrical device, some electrical energy is converted to heat energy by resistance. This heat energy is lost to the mechanical or electrical system. Therefore, to have perpetual motion, this lost energy would have to be replaced by energy created in the system itself. The law of conservation of energy ("Energy can be neither created nor destroyed.") says that this is impossible. If the voltage generated by moving a conductor through a magnetic field were of such a direction as to *aid* the motion (by electromagnetic action), you would have a perpetual-motion device, which the argument above has just proved to be impossible. Were it not impossible, your car's alternator, once started, could run your car *and* generate electricity to charge the battery; the engine would be unnecessary. The principle of countervoltage generation and polarity are compactly stated in Lenz's law: *When a current results from a magnetically induced voltage, its direction is such that its magnetic effect opposes the motion causing the induction.*

SUMMARY

1. A magnetic field is a region in which ferromagnetic materials and current-carrying conductors experience a force without physical contact.

2. Every magnet has a north and south pole. Like poles repel; unlike poles attract.

3. Flux density B = lines of flux per unit area.

4. A moving charge produces an electromagnetic field, in accordance with the left-hand rule. The left-hand rule also works for coils. The magnetic effect of a current in a coil is the "magnetomotive force" $\mathcal{F}$, which equals the number of turns $\times$ the current.

5. Ferromagnetic materials concentrate fields in their vicinity by the factor μ (permeability). The magnetic effect of a particular piece of material is its permeance $\mathcal{P}$. $1/\mathcal{P}$ = reluctance $\mathcal{R}$.

6. Retentivity refers to a magnetic material's ability to stay magnetized.

7. Demagnetization can result from heat, shock, or strong fields.

8. The magnetic field produced by a moving charge will result in a force on that charge if the path of the charge is through another magnetic field.

9. A voltage is induced in a conductor that moves with respect to a magnetic field. The polarity of that voltage is such as to produce a current whose magnetic effect opposes the motion that caused the induced voltage.

QUESTIONS

1. Define:
   (a) Magnetic field
   (b) Magnetic poles
2. What is the law of attraction and repulsion?
3. What does each of these symbols stand for?
   (a) B
   (b) $\mathcal{F}$
   (c) μ
   (d) $\mathcal{P}$
   (e) $\mathcal{R}$
   (f) ϕ
4. What is retentivity?
5. What are three ways by which a permanent magnet can lose its magnetism?
6. Describe the motor effect.
7. Describe the generator effect.
8. What is countervoltage?
9. What is the direction of a current that results from an electromagnetically induced voltage?

3

RELAYS

A relay is basically a remotely controlled switch. It is probably the simplest electromechanical device, yet it is one of the most important components of many industrial electrical systems. As shown in Fig. 3-1, the relay is simply a set of switch contacts that are opened or closed by an electromagnet.

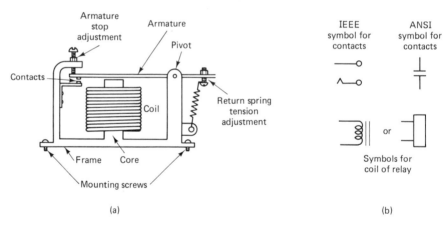

Figure 3-1 SPST relays: (a) parts of a relay; (b) circuit symbols for SPST relay.

RELAY COILS

For any relay coil, there are two critical voltages. The first is the minimum voltage required to cause the electromagnet to "pull in" the armature, operating the relay. This voltage is sometimes called the pull-in voltage, but more properly, the *operate voltage*. The other critical voltage is the *release voltage*, that is, the voltage at which the relay, once operated, will release. The two voltages are not the same. For a typical relay whose coil is rated at 12 V, the operate voltage might be around 10.5 V, and the release voltage, around 8 V. A moment's thought will show why these voltages differ. Remember $\phi = \mathfrak{F}/\mathfrak{R}$? As shown in Fig. 3-2, the magnetic circuit of an operated relay contains a much smaller reluctance than that of a released relay. This means that the same mmf will produce more flux in the closed relay. But the spring tension trying to open the armature is roughly the same whether the relay is operated or released. Thus the greater flux available in the operated position is not needed, and the mmf can be reduced without the relay releasing. Since $\mathfrak{F} = N \times I$, reducing $\mathfrak{F}$ means reducing I. The coil resistance is constant, so less current means less applied voltage. The difference between the operate and release voltage is called *backlash* (or sometimes, *hysteresis*, not to be confused with magnetic hysteresis).

The third characteristic of a relay coil is the *operate current*. Sometimes the resistance of the coil is given instead; Ohm's law tells you that if you know the operate voltage and either the current or the resistance, you can calculate the missing quantity. The coil characteristics are illustrated in Fig. 3-3. Table 3-1 shows the operate and release voltages and operate current for several commercially available relays.

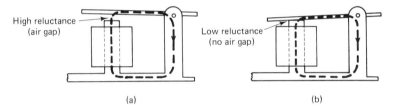

(a)   (b)

Figure 3-2 Magnetic circuits of operated and released relays: (a) released; (b) operated.

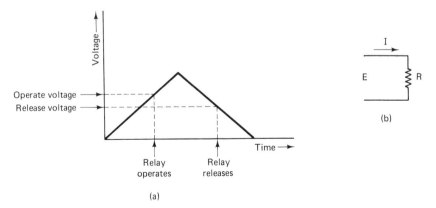

Figure 3-3 Relay coil characteristics: (a) operate and release voltages; (b) coil current and resistance.

TABLE 3-1 Coil Characteristics of Typical Relays

Description	Coil rating		Operate voltage	Release voltage	Operate current
	Volts	Amperes			
Telephone type	48 dc	6 mA	32	22	4 mA
Standard clapper	230 ac		170 ac, 106 dc	130 ac, 15 dc	12.9 mA ac, 13.3 mA dc
Standard clapper	12 dc	85 mA	8.5	2.7	52 mA
Miniature clapper	24 dc	40 mA	20	14	33.5 mA
Miniature clapper	12 dc	67 mA	9.4	7.0	30.2 mA
Low-current clapper	12 dc	10 mA	9.6	5.0	8 mA
Low-current clapper	6 dc	11 mA	4.3	1.8	7.9 mA

RELAY CONTACTS

The switching in a relay is accomplished by the contacts. These can be made of any conducting material. The least expensive contact material in common use is copper. Often, though, contacts are made of tungsten, silver-plated copper, or gold. The usual reason these particular materials are used is that they resist oxidation well. When switch or relay contacts are in the process of opening or closing, a spark is likely to result. This can be a tiny spark or a severe spark, depending on the amount of current being switched and the type of circuit. Sparks tend to oxidize (burn or rust) contacts. Tungsten is a tough, nonoxidizing metal that is used for general-purpose relays. Silver and gold are softer and have lower resistance, and are used in circuits in which a good connection is more critical, such as very low level audio circuits and high-current circuits. Obviously, they are more expensive than tungsten, but their higher reliability more than offsets the extra cost for these applications.

Relay contacts also vary in their surface area. High-current relays require a large contact area, whereas lower-current relays use smaller contacts.

The prime enemy of relay contacts is contamination: dirt, grease, and grime! Most relay contacts are exposed to the air, where they naturally pick up contaminants over a period of time. The effects of contamination are reduced by the wiping action provided in most relay contacts. That is, when the contacts close or open, the two contacts are made to slide across each other somewhat, wiping the contamination off each other's surfaces.

The simplest electrical form that relay contacts can have is that of an SPST (single-pole, single-throw) switch. This is the form that either opens or closes a single circuit. There are several varieties of SPST relays, as diagrammed in Fig. 3-4. The names require a little explanation. If a relay closes a circuit when it is operated, it is called a *normally open* relay (or *make-contact relay*). This is because the contacts are open when the relay is in its

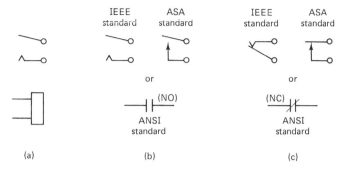

Figure 3-4 SPST relay contact symbols: (a) normal contact position; (b) normally open SPST contact symbols; (c) normally closed SPST contact symbols.

normal (i.e., nonoperated) condition. A relay that opens a circuit when operated is called a *normally closed* (or *break-contact*) relay. Notice that schematic drawings always show relay contacts in the normal (released) position.

A relay can be made with virtually any number of moving contacts. The number of moving contacts (i.e., contacts attached to the armature) is called the number of *poles* of the relay. A normal relay can only have two *throws*, or fixed contacts per pole. (60PDT relays are commonly available). Thus a relay that has a single contact arm attached to the armature, with one fixed contact being closed when the relay is not operated and another being closed when the relay is operated, would be a single-pole, double-throw (SPDT) relay. Schematics for a variety of relay types are shown in Fig. 3-5. For any double-pole relay, there are two possible variations. If the moving contact stops touching fixed contact A before it begins to touch fixed contact B, it is referred to as a *break-before-make* or *nonshorting* contact. If, instead, there is a brief instant during the switching action when the moving contact is touching both fixed contact A and fixed contact B, the relay is a *make-*

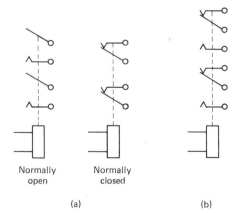

Normally
open

Normally
closed

(a)   (b)

Figure 3-5 Double-pole relay contacts: (a) DPST; (b) DPDT.

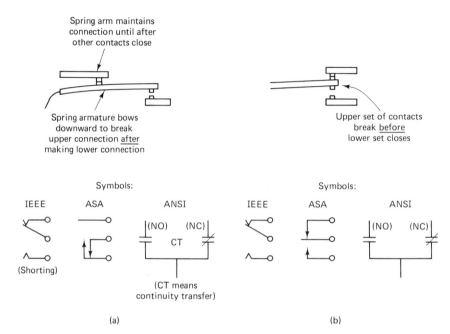

Figure 3-6 Shorting and nonshorting relays: (a) shorting type; (b) nonshorting type.

before-break or *shorting-type* relay (see Fig. 3-6). For most applications, nonshorting contacts are used. However, for certain applications in audio or metering circuits, an objectionable thump or hum can result if a circuit is momentarily opened. For these applications, shorting-type contacts are often used.

PHYSICAL CONSTRUCTION

There are several different ways in which relays are usually built. The simplest, cheapest, and therefore most common is the *clapper-type* relay (Fig. 3-7a). For particularly high current applications, clapper-type relays may be supplied with contacts in which the circuit is made or broken through a drop of mercury (Fig. 3-7b). These *mercury-wetted* contacts resist sparking better than any other type. Where many low-current circuits must be switched by a single relay, small contact parts and very good wiping action are needed. These are provided by the *telephone-type* relay (Fig. 3-7c). If a relay needs to be quite small or have very fast switching time with extremely good reliability, a *reed relay* can be used (Fig. 3-7d). The reed relay operates on somewhat different principles than do other relays. It contains two reeds made of a ferromagnetic metal. In the presence of a magnetic field, these reeds pull

Figure 3-7 Four types of relays: (a) clapper (courtesy of Price Electric Co.); (b) mercury-wetted (courtesy of Clare Div., General Instrument Corp.); (c) telephone; (d) reed (courtesy of Magnecraft Electric Co.).

together or apart, depending on whether the relay is a normally-open or a normally-closed type. The reeds themselves have very low mass, providing fast switching times. They are encased in an evacuated or a nitrogen-filled glass tube, so there is no oxygen to cause oxidation, and no contaminants can enter. Thus their lifetimes are numbered in the millions of switching cycles.

There are some applications for which a relay is needed that will select one of a large number of possible connections for a single moving contact. An example would be a telemetry (remote-metering) circuit, as diagrammed in Fig. 3-8. The type of relay used for this purpose is called a *stepping*, or *stepper* relay. It is a relay version of a rotary switch. In the figure, stepping relay X_1 advances one position each time switch SW is pressed, allowing a voltmeter to measure any one of six different circuits, using only two telephone lines. Often a second stepping relay, synchronized with X_1, would be used at the metering location to light lamps indicating which circuit is being measured.

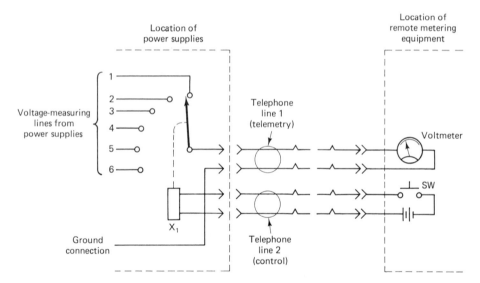

Figure 3-8   Use of stepper relay in telemetry circuit.

Stepper relays can be made with virtually any number of poles and virtually any number of throws. Four-pole, 32-position (4P32T) stepping relays are not uncommon. Figure 3-9 shows a stepping relay and its schematic. Stepping relays have no "normal" position. They are activated by current pulses. Each pulse moves the contacts to the next position. Between pulses, the relay remains in its last position. Some stepping relays have mechanisms that can rotate the relay in either direction, depending on which of the two

(a)

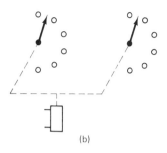

(b)

Figure 3-9 Stepping relay:
(a) two types of stepping
relays (courtesy of Guardian
Electric Co.); (b) schematic
(6PDT).

terminals is pulsed with current. These are called bidirectional stepping relays.

Another relay that is activated by pulses is the *sequencing relay*. This relay also has no "normal" position. It has two or more stable positions, and changes from one to another in a predetermined sequence when the coil is pulsed. Figure 3-10a shows a sequencing relay. The coil of this particular relay operates a ratchet wheel and a control cam that cause the contacts to change connections.

Related to the sequencing relay is the *mechanical latching relay*. Such a relay is activated by a pulse, and then remains activated even after the pulse is removed. Single-coil latching relays are deactivated by a second pulse; whereas dual-coil units are deactivated by a pulse to the second coil, which is called the release coil. Both types of latching relays are pictured in Fig. 3-10b.

Sometimes a relay is needed that will change its connections at some fixed amount of time after the control voltage to the coil is changed. Such relays are called *time-delay relays*. They come in two varieties. The first has a delay between application of coil voltage and operating. This type is called a *delay-on-activate* (DOA) time-delay relay. An application would be a burglar alarm control that gives the owner 30 seconds after turning the alarm system on before the system is activated. The owner can leave the protected area and lock the doors during those 30 seconds, without tripping the alarm. The second variety of time-delay relay is one that remains operated for a while after the coil voltage is removed. This is called a *delay-on-release* (DOR) relay. An example of its use would be in the early cars whose headlights remained on for a minute or so after the switch was turned off to light one's way to the door. The newer cars do this electronically, but the older ones used a time-delay relay. An example of a mechanical time-delay relay is shown in Fig. 3-10c.

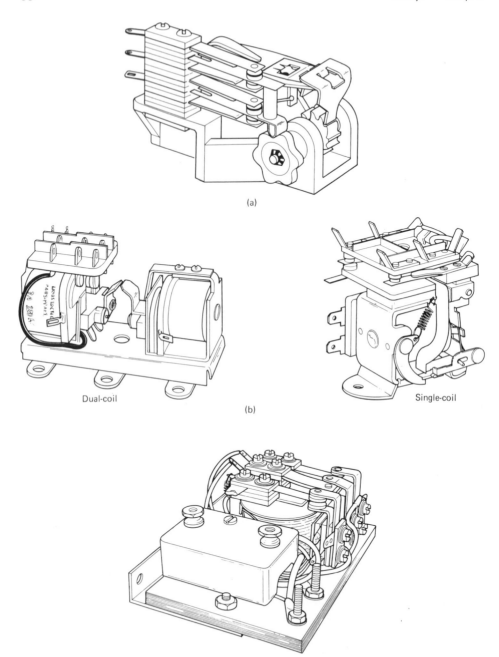

(a)

Dual-coil

(b)

Single-coil

(c)

Figure 3-10 Sequencing, mechanical latching, and time-delay relays: (a) sequencing relay; (b) mechanical latching relays; (c) time-delay relay. (Courtesy of Relay Specialties, Inc.)

As you would expect, the inexorable march of progress has its effect on relay design. Many relays are now being replaced by solid-state relays that are entirely electronic and have no moving parts. Externally, these usually look exactly like enclosed mechanical relays. As you would expect, solid-state relays require no maintenance. However, when a solid-state relay fails, repair is usually not feasible; the unit must be replaced.

APPLICATION OF RELAYS

The most common use for relays is to open or close high-current circuits without having long wire runs. Figure 3-11 illustrates this sort of application. Another typical example would be the starter relay in an automobile. The coil of this relay is activated by the ignition switch. There are several feet of wire between the ignition switch and the relay. The current is less than an ampere, so a large-diameter wire is not required to keep the voltage drop low. When the relay is operated, it closes a circuit to the starter motor, which can draw 40 or more amperes. The maximum resistance in the total circuit for a current of 40 A at 12 V is 0.3 Ω. Most of this resistance is in the starter motor itself. To minimize the length of the high-current leads, the relay is mounted in the starter-motor case.

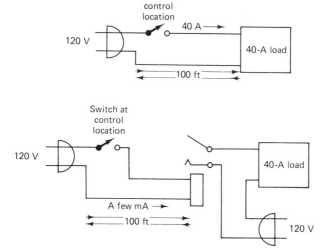

Figure 3-11   Use of a relay to control a high-current load.

Sometimes relays must be activated by a pulse of current, and must remain activated after the pulse is removed. The sequencing and latching relays mentioned earlier can do this job, but more commonly a standard double-pole relay in a *latching circuit* is used. The latching circuit is shown in Fig. 3-12. It operates as follows:

1. The switch is pressed momentarily, operating the relay.

2. The contacts close, providing a path for current from the supply through the relay contacts to the coil, so the relay "latches" closed.

3. To release the relay, the power must be disconnected. This can be done by a momentary-contact, normally closed pushbutton switch.

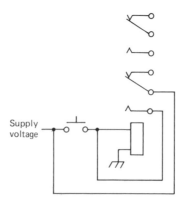

Figure 3-12   Latching connection for relay.

A typical example of this sort of application would be the relays that are often used to operate saws, planers, or sanders in woodworking shops. These have a pushbutton ON switch and a pushbutton OFF switch. When the ON switch is pressed, a current pulse is sent to the relay coil, activating the relay, which then latches. The relay remains activated until the OFF button is pressed, breaking the circuit to the relay coil. (If a sequencing relay had been used, the same button would have to be used for both ON and OFF, which could confuse the operator of the machinery.) Another common application of a latching relay is in the alarm sounder circuit of a burglar alarm. When a normally open alarm circuit is momentarily closed, the resulting current pulse closes a relay that feeds power to the sounder (bell or siren). The relay latches closed so that the sounder continues to operate until someone manually resets the control panel, breaking the connection to the relay coil.

A very important circuit in which relays are used in industrial equipment is the sequence circuit. This is the type of circuit that operates automatic or semiautomatic equipment that performs a number of different functions in a certain sequence. Figure 3-13 illustrates such a circuit. Switches A, B, and C in the figure could represent either manual switches under the control of a human operator (semiautomatic machine) or sensor switches in the machine that signal the completion of one step so that the next step can begin (automatic machine). The operating sequence is as follows:

1. When SW_A is pressed, relay X_1 is activated.

2. Power is supplied to relay X_2 through the contacts of X_1. When SW_B is pressed, a ground is provided, and X_2 is activated.

3. Power to relay X_3 comes through the contacts of X_2. When SW_C is pressed, a ground is provided, and X_3 is activated.

Notice that the relays must operate in sequence: X_2 cannot operate unless X_1 is operated, and so on.

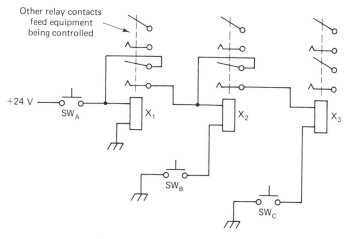

Figure 3-13 Sequence circuit.

Another common application for relays is to open or close a circuit when the current or voltage reaches a certain level. For this type of application, the operate and release voltages of the relay are critical. Some relays for this application are made so that the operate and release voltages are nearly the same (very little backlash); these are called *sensitive relays*. The voltage regulator relay in older automobiles is such a relay. Figure 3-14 shows cir-

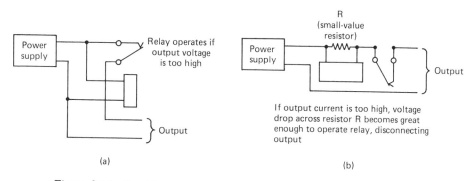

Figure 3-14 Sensitive-relay applications: (a) overvoltage protection; (b) overcurrent protection.

cuits in which the relay operation is triggered by the circuit's voltage or current level.

RELAY DRIVER CIRCUITS

Many, if not most relays in use today are driven by electronic circuits. There are several reasons for this. It is true that a relay can switch many amperes with a control (coil) current of a few milliamperes. However, sometimes a much smaller control current—only a few microamperes—is all that is available. Electronic devices can amplify this control current to operate the relay. Second, many sensors in automatic and semiautomatic equipment produce small changes that must be converted by electronic devices into control currents to drive relays. For example, a temperature sensor might respond to temperature variations by changing its resistance. At a certain temperature, the resistance would reach a critical level that would cause an electronic circuit to send a control current to operate a relay. Third, more and more equipment is being built in which the various production steps are controlled by a microcomputer or some other form of electronic logic circuit. These circuits may then control the operation of motors, hydraulic valves, and so on, via relays. There are four electronic devices in common use in relay-driving circuits. We will take a look at how each of these devices works.

The first electronic device invented was the vacuum tube. The theory behind vacuum-tube operation is beyond the scope of our study. However, we can look briefly at the way a vacuum tube operates as a switch. As shown in Fig. 3-15, a vacuum tube can have various forms of internal construction. A two-element tube is called a diode; a three-element one, a triode; a four-element one, a tetrode; and a five-element one, a pentode. Only triodes and pentodes are of concern to us now. In both of these types of tubes, there is a *cathode*, a *control grid*, and a *plate* (sometimes called the *anode*). In normal operation the plate has a voltage applied that makes it from 50 to 500 V more positive than the cathode. The resulting current through the tube is controlled by the voltage between control grid and cathode. If this voltage is zero, a large current can flow. For this reason a vacuum tube is called a "normally on" device. If the control grid is a few volts negative compared to the cathode, less current can flow. If the control grid is made many volts nega-

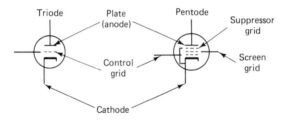

Figure 3-15 Vacuum-tube schematic symbols.

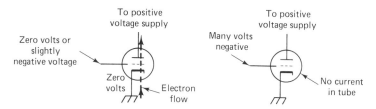

Figure 3-16   Current flow in vacuum tube.

tive, the tube is "turned off" and no current flows. This is illustrated in Fig. 3-16. A vacuum tube can thus be considered a variable resistor whose resistance is determined by the voltage between control grid and cathode. For our purposes, we can consider a tube to be a switch that is turned on by having no voltage applied to the control grid, and turned off by having the grid made 20 V or so more negative than the cathode. No current is drawn by the control grid; tubes are voltage-controlled devices. If the tube is to be used to control a relay, the relay can be put either in the plate circuit or in the cathode circuit. Figure 3-17 illustrates these two ways of using a vacuum tube to control a relay. Notice that either of these circuits requires a control voltage of either zero or −20 V, with a current in the picoampere range. Yet the circuits control a 120-V 10-mA relay coil.

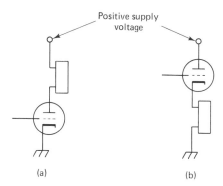

(a)

(b)

Figure 3-17   Relay coils controlled by vacuum tubes: (a) relay in plate circuit; (b) relay in cathode circuit.

Because of their fragility, relatively large size, and tendency to burn out, vacuum tubes started disappearing from circuit designs as soon as something better became available. That "something better" was the bipolar transistor. It was invented in the 1940s, but didn't come into common use until the production costs became reasonable—in the 1960s. Like a vacuum tube, a transistor also has three terminals. These are called the *emitter*, *base*, and *collector* (see Fig. 3-18a and b). Bipolar transistors come in two *polarities*, NPN and PNP. The operation is the same in either, except that the polarity of the applied voltages is opposite. This is illustrated in Fig. 3-18c and d. We

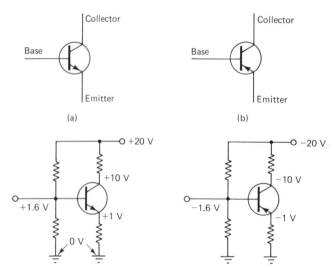

Figure 3-18 Bipolar transistor schematic symbols and voltages: (a) NPN symbol; (b) PNP symbol; (c) NPN circuit; (d) PNP circuit.

will limit our discussion to NPN transistors for simplicity. Just remember that if every positive voltage were changed to negative, the discussion could be applied to PNP transistors.

In normal operation of an NPN transistor, a voltage is applied that makes the collector more positive than the emitter. If no voltage is applied to the base, no current will flow. This means that the transistor is a "normally off" device. To make the transistor conduct, the base must be made about six-tenths of a volt more positive than the emitter. Then the amount of current that flows in the transistor is controlled by the amount of current flowing from the base terminal. This base current is amplified by the transistor's current gain, or β (Greek lowercase letter beta). Thus a transistor having a β of 100 would have 100 mA of emitter-to-collector current for each milliampere of base current. For transistors used as relay drivers, we can think of the transistor as a switch that is turned off if no base voltage is applied, and turned on if the base is made at least six-tenths of a volt more positive than the emitter. (The base circuit must also have low enough resistance so that sufficient base current can flow. The circuits that supply this base current are outside the scope of our present study.) If the transistor is to be used to control a relay, the relay can be put in either the emitter circuit or the collector circuit. These two configurations are shown in Fig. 3-19.

There is another type of transistor, one that has become much more common in the last decade: the *field-effect transistor*, or *FET*. These devices come in two varieties. *Depletion-mode FETs* are normally on, voltage-operated devices; in other words, they behave almost exactly like vacuum tubes. *Enhancement-mode FETs* are normally off, voltage-operated devices. In other words, they behave like transistors except that they are controlled by voltage instead of current. The three terminals of a FET are called the

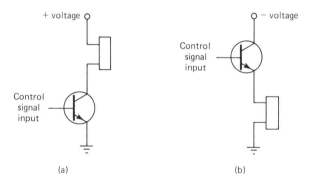

Figure 3-19 Relay coils controlled by transistors: (a) relay in collector circuit; (b) relay in emitter circuit.

source, *drain*, and *gate*. These correspond directly to the cathode, plate, and grid of a vacuum tube. The only type of FET that is used very often in connection with electromagnetic devices is the power FET, sometimes called the VFET or HexFET.* They look exactly like bipolar transistors, but operate with different bias voltages, as shown in Fig. 3-20.

There are two devices that act somewhat like transistors in switching circuits, but "latch on" once they begin to conduct. These are the *SCR* (sili-

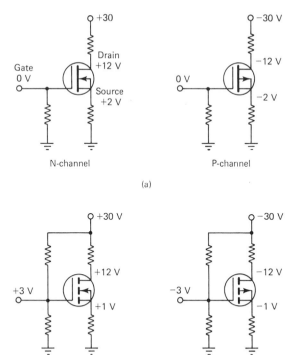

Figure 3-20 Field-effect transistor schematic symbols and voltages: (a) depletion-mode FETs; (b) enhancement-mode FETs.

*HexFET is a registered trademark of International Rectifier Corp.

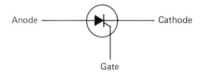

Figure 3-21   SCR schematic symbol.

con-controlled rectifier) and the *triac*. We will discuss the SCR first. An SCR has three terminals, called the *anode*, *cathode*, and *gate*. The schematic symbol is shown in Fig. 3-21. In normal operation, a voltage is applied that makes the anode positive with respect to the cathode. With no gate voltage, no current flows. Thus the SCR is a normally off device. If the gate is made more positive than the cathode, then cathode-to-anode current will flow. The current is determined by whatever is in series with the SCR and the power supply. At this point we can see two differences between the SCR and the transistor. First, the gate current of an SCR has no control over the cathode-to-anode current, whereas the base current of a transistor does control emitter-to-collector current. Second, if we now remove the control voltage that we applied to the SCR's gate, the SCR continues to conduct. A transistor with no base-to-emitter voltage will not conduct. This is why we

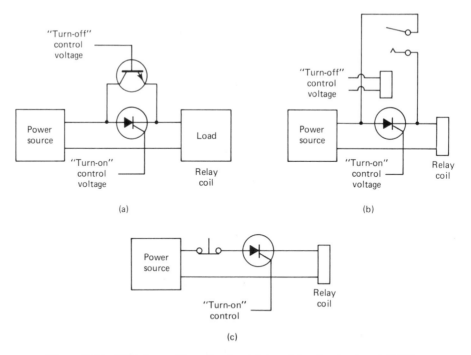

Figure 3-22   SCR turn-off methods: (a) transistor to short out SCR's anode-to-cathode voltage; (b) relay to short out SCR's anode-to-cathode voltage; (c) normally closed switch to momentarily interrupt supply voltage to SCR.

said that the SCR will "latch." It acts rather like a latching relay. To turn off the SCR, it is necessary to disconnect or short out the anode-to-cathode voltage. This can be done in several ways, as shown in Fig. 3-22.

The SCR, like a transistor, conducts current in only one direction. Sometimes a device is needed that will conduct current in either direction, that is, a device that will conduct ac. Such a device is called a bidirectional SCR or a *triac*. A triac's three terminals are called *anode 1*, *anode 2*, and *gate*. A triac will conduct if:

1. Anode 2 is more positive than anode 1, and the gate is more positive or more negative than anode 1, or if

2. Anode 2 is more negative than anode 1, and the gate is more positive or more negative than anode 1.

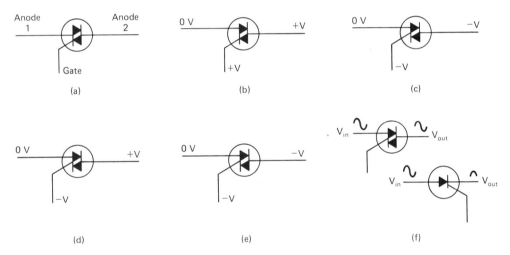

Figure 3-23 Conduction in triacs: (a) symbol; (b) triac conducts when gate and anode 2 are both more positive than anode 1; (c) triac conducts when gate and anode 2 are both more negative than anode 1; (d) triac conducts when gate is more negative and anode 2 is more positive than anode 1; (e) triac conducts when gate is more positive and anode 2 is more negative than anode 1; (f) once conducting, triac passes ac; SCR rectifies ac.

These conditions are shown in Fig. 3-23. Like an SCR, a triac, once conducting, remains conducting until the anode-1-to-anode-2 voltage is disconnected or shorted out. (If the anode-to-anode voltage is reversed in polarity, the triac will turn off as the voltage passes through the zero point.) Circuits using SCRs or triacs to control relays are shown in Fig. 3-24.

Sometimes relays are used in circuits in which the *operate time* (time required for contacts to make or break after coil is energized) is critical. In such circuits, the operate time may be controlled by a capacitor connected

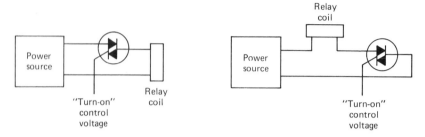

Figure 3-24   Methods of controlling relay using SCR or triac.

in parallel with the relay coil. Then, when voltage is applied to the coil, the capacitor must charge up to the relay's operate voltage before the relay will operate. The operate time is then given approximately by

$$T_O = R_{\text{coil}} \times C$$

The larger the capacitor, the slower the operation. The release time is similarly increased, since the capacitor must discharge to the relay's release voltage before the relay can release.

TROUBLESHOOTING, MAINTENANCE, AND REPAIR OF RELAY CIRCUITS

By far the most common problem that occurs in relay circuits is plain ol' dirt. Even when relay contacts are surrounded by plastic or metal dust covers, a certain amount of airborne oil and dust manages to make its way to those contacts. Since most grime (*contaminants*, in more formal language) is electrically insulating, a relay with dirty contacts may operate perfectly but still fail to complete the circuit. The remedy for dirty contacts is clean-·ing with a good quality spray-type relay cleaner. The cleaner should be sprayed directly on the contacts and the contacts should be opened and closed several times in rapid succession. This procedure will aid the natural wiping action of the contacts in pushing the dirt off the surfaces where the electrical connection is made. In some cases, a cleaner-saturated cloth will have to be pulled through between the closed contacts to clean away particularly stubborn contaminants. Figure 3-25 shows these methods of cleaning relay contacts.

Relays that are used in high-current applications are sometimes subject to significant sparking at the contacts. This sparking tends to build up a layer of oxide on the surface of the contacts. Oxide, like dirt, is usually insulating, so it prevents the contacts from making a good connection. Oxidized contacts can sometimes be cleaned by the solvent-saturated-cloth method. Often, though, tougher measures are necessary. If the contacts are not plated

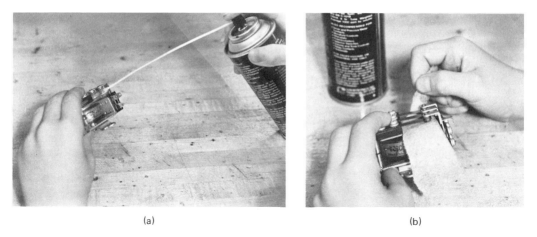

(a)   (b)

Figure 3-25 Methods of cleaning relays: (a) spraying with cleaner; (b) using solvent-saturated cloth.

(i.e., if they are solid copper or tungsten), they can be cleaned by pulling a piece of fine emery paper through the closed contacts. In extreme cases, where the sparking has pitted or otherwise deformed the contacts, a relay burnishing tool can be used. This tool is a very fine file that reshapes the contact surfaces. It is essential to be very careful in using a burnishing tool so that the proper contact shape is maintained (see Fig. 3-26). Relays with plated contacts have to be replaced if the contacts oxidize.

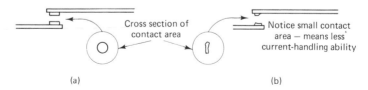

(a)   (b)

Figure 3-26 Proper shape of relay contacts: (a) properly formed; (b) improperly formed (result of careless filing of contacts or severe pitting from sparking).

Every now and then a relay fails to operate mechanically. This can be a fault of the driving circuit such as a defective tube or transistor. It can be caused by an open relay coil, although this is very rare. If the coil is not open and the proper operating voltage is present, the relay can be adjusted by means of the adjusting screws or tabs. Figure 3-27 shows how this is done. The armature stop adjustment affects the operate voltage. The return spring adjustment affects both the operate and the release voltage. Thus if only the operate voltage needs to be adjusted, the armature stop adjustment would be changed. If both voltages need adjustment, the return spring tension would

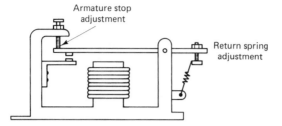

Figure 3-27 Relay adjustments.

be adjusted first until the release voltage was correct; the armature stop could be adjusted if necessary to provide the correct operate voltage. If no adjustment screw is provided, the armature stop and return spring tabs can be adjusted by bending with pliers. In setting the armature stop, it is very important that the gap between contacts be kept large enough so that the normal circuit voltages do not arc or spark across the contacts when they are open. This means a minimum opening of about 0.01 in. for each 30 V. Thus an opening of at least 0.033 in. (33 mils) would be required for a relay that was switching 100 V. For relays that feed motors, large electromagnets, or other inductive loads, the gap should be several times larger. A feeler gauge can be used to check the opening. Usually, any adjustments that have to be made on a relay are very small, so maintaining the proper opening is usually not a problem.

In circuits that switch dc, it is often possible to reduce sparking by installing a diode connected across the relay contacts. The diode must be installed so that it blocks the flow of the direct current. When the relay switches, any sparking represents "spikes" of ac superimposed on the dc. The diode shorts out one-half of each spark cycle, thus reducing the number of actual arcs. A small capacitor (0.01 to 0.1 μF) will also work in many cases, and can also be used in some ac circuits. The voltage rating of such a diode or capacitor should be at least five times the peak value of the voltage that the relay is switching. Figure 3-28 illustrates these methods of reducing sparking.

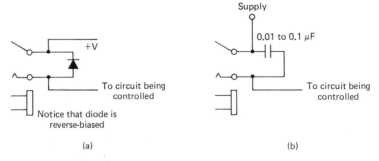

Figure 3-28 Methods of reducing arcing of contacts: (a) diode (dc circuits only); (b) small capacitor (dc or ac circuits).

SOLENOIDS AND RELATED DEVICES

Many industrial devices are activated by specialized electromagnets that closely resemble relay coils. These are called *solenoids*. A solenoid has a rated voltage, an operate voltage, and a release voltage, as does a relay. Since the solenoid does not open or close an electrical circuit, there are no contact ratings. But there are two other important ratings: how far the core (or *plunger*) moves, and how hard it pulls. Most solenoids are linear types, as shown in Fig. 3-29a. But some applications require a rotary solenoid, of the type shown in Fig. 3-29b. A linear solenoid will be rated in terms of inches or centimeters of travel (*throw*) and ounces, pounds, grams, or kilograms of force. Rotary solenoids are rated in degrees of rotation and inch-ounces or centimeter-grams of torque. Solenoids seldom require service. Occasionally, though, they will become "hung up" in either the operate or the released condition. The usual cure is to spray some contact cleaner on the plunger to dissolve the stray dirt or gum that is causing it to stick. In severe cases, it

(a)

Figure 3-29 Linear and
rotary solenoids: (a) linear;
(b) rotary. (Courtesy of IMC
Magnetics Corp and Ledex,
Inc.)

(b)

may be necessary to disassemble the solenoid and clean the plunger (not the coil!) in Varsol or lacquer thinner. (Solvents such as lacquer thinner will attack the insulation on the coil's wire.) The usual reason that such measures would be required is an earlier decision by some uninformed person that the solenoid should be oiled. Over a period of time, oil can collect grease and become gummy. With sufficient heat and aging, oil can turn into a varnish-like material. The best prevention is to remember that solenoids essentially never need lubrication. If a solenoid must be lubricated, silicon lubricant, graphite, or a spray Teflon* powder should be used.

A specialized solenoid-operated device is the solenoid valve. These are used to control the flow of fluids or gases by means of an electrical signal. Solenoid valves are rated in terms of operate voltage, release voltage, maximum pressure, inlet and outlet port size, and whether they are gastight or not. Figure 3-30 shows two solenoid valves.

Figure 3-30   Solenoid valves.

SUMMARY

1. A relay is a remote-controlled switch.
2. Relay coils are specified as follows:
   a. Rated voltage
   b. Operate voltage
   c. Release voltage
   d. Coil current or resistance
   (Specifications b and c are often not given by the manufacturer.)
3. Relay contacts come in many forms; the number of moving contacts is called the number of poles; the number of fixed contacts per pole is called the number of throws. Contacts are specified as follows:
   a. __P __T (number of poles and throws)
   b. Rated current
   c. Rated voltage
4. SPST relays may be normally open or normally closed.

*Teflon is a registered trademark of E.I. DuPont de Nemours & Co.

5. Double-pole relays may be make-before-break (also called shorting), or break-before-make (also called nonshorting).

6. Relays are available in clapper-type, telephone-type, mercury-wetted, or reed construction, for various specific applications.

7. A stepping relay is a remote-controlled rotary switch.

8. Sequencing relays have two stable positions.

9. Time-delay relays may offer a delay before activating or a delay before releasing, after the coil control voltage is changed.

10. Relays can be made to remain activated after the control voltage is removed by means of a latching circuit.

11. Relays in semiautomatic or automatic machinery are often connected in sequence circuits. Such circuits perform a series of operations in a predetermined sequence.

12. Sensitive relays have tightly controlled operate and release voltages.

13. Relay control voltages may be derived from vacuum tube, transistor, SCR, or triac circuits.

14. Relay operate time can be modified by the use of capacitors.

15. Most relay maintenance consists of cleaning contacts, although occasional adjustments may be required.

16. Contact sparking can be reduced by the use of diodes or capacitors.

17. Small electromagnets called solenoids perform many functions in industrial equipment.

18. Solenoids are rated in terms of throw and force or pull.

19. Some valves used in industrial equipment are operated by integral solenoids.

QUESTIONS

1. What is the operate voltage of a relay? the release voltage? Which is higher?

2. What is the purpose of using silver and gold in relay contacts?

3. Draw the diagram for a 6PDT relay.

4. How is a shorting-type contact different from a nonshorting type? What is a common application for relays having shorting-type contacts?

5. Name the major advantages of:
   (a) Mercury-wetted relays
   (b) Reed relays

6. What kind of relay could have DP24T contacts?

7. What two types of time-delay relay are available?

8. Draw a diagram of a latching connection for a DPDT relay.

9. Draw a diagram of a three-step sequence circuit. Explain its operation.

10. What is a sensitive relay? What is it used for?

11. Describe the operation of a vacuum-tube relay driver. Include typical voltages for the circuit.

12. Draw a diagram and explain the operation of:
    (a) A transistor relay-driver circuit
    (b) An SCR relay-driver circuit
    (c) A triac relay-driver circuit

13. What is required to turn on:
    (a) A vacuum tube?
    (b) A transistor?
    (c) An SCR?
    (d) A triac?

14. What is the difference between PNP and NPN transistors, as far as their operation is concerned?

15. Describe three methods of cleaning relay contacts and when each would be used.

16. How do you know whether a relay needs adjusting? Describe the adjustment procedure.

17. Why is a diode or small capacitor sometimes connected across relay contacts?

18. What is a solenoid?

19. How do you service a solenoid that has gotten "hung up"?

4

GENERATORS

GENERATOR BASICS

A generator is a device that converts mechanical energy (usually rotational energy) into electrical energy. The principle that makes a generator work was introduced in Chapter 2: When there is relative motion between a conductor and a magnetic field, a voltage is created (induced) in the conductor. There are, then, two necessary parts of any generator. The *field* is the part of a generator that supplies the magnetic field, and the *armature* is the conductor in which the output current is generated (see Fig. 4-1).

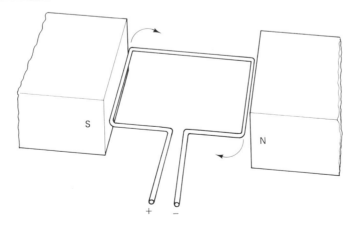

Figure 4-1 Basic generator.

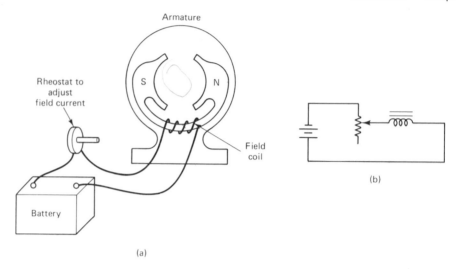

Figure 4-2 Generator with externally excited field: (a) pictorial; (b) schematic.

The field can be either a permanent magnet or an electromagnet. Most often it is an electromagnet. The current for the electromagnet can be supplied by an external battery. Figure 4-2 shows a generator having an externally excited electromagnetic field. Other ways of obtaining current for the field coil will be discussed later in this chapter.

The problem with the basic generator shown in Fig. 4-1 is that after the armature rotates several times, the leads will have a serious twisting problem.

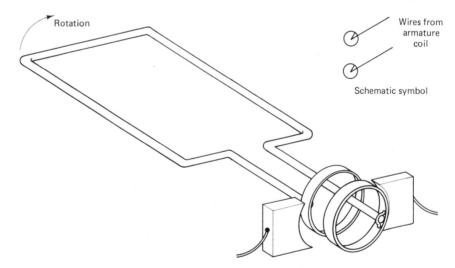

Figure 4-3 Slip rings and brushes.

Naturally, this problem was discovered very early in the experimental work on generators, and a solution was found. The solution was to make connection to the armature by means of sliding contacts, called *slip rings* (see Fig. 4-3). Connection to the slip rings is made by means of carbon or copper blocks called *brushes*.

GENERATOR OUTPUT WAVEFORMS

It was mentioned in Chapter 2 that the faster a conductor moves, relative to a magnetic field, the more voltage is induced. Another way of saying this is: the more lines of force a conductor crosses each second, the greater is the induced voltage. A flat coil rotating at one revolution each 10 seconds in a magnetic field does not cross the same number of lines of force each second. Figure 4-4 shows why this is so. In Fig. 4-4a, as the armature rotates from $0°$ to $10°$, few lines of force are crossed. Hence very little voltage is induced. (Notice that the "x" end of the armature is moving downward through the lines of force.) In Fig. 4-4b, as the armature rotates from $80°$ to $90°$, many lines of force are crossed, so a large voltage is induced. As the armature passes $90°$, the induced voltage is maximum. In Fig. 4-4c, moving from $170°$ to $180°$, the armature again crosses a minimum number of lines of force, resulting in little induced voltage. After the armature passes $180°$ (zero voltage), the amount of induced voltage again begins to increase, but since the "x" end of the armature is now moving upward through the lines of force—its direction of motion is reversed—the polarity of the output voltage is also reversed.

Since a given section of the coil does not cross the lines of force in the same direction each time, the result is an induced voltage that alternates, or changes direction, each half-revolution. Thus a generator of this type produces an ac output.

Few ac generators—or alternators, as they are usually called—are built with only one armature coil like the one shown in Fig. 4-4. Usually, there are three coils mounted at $60°$ angles from each other. Some alternators have six or more armature coils. The number of separate windings of an armature is usually given in terms of the number of *poles*. A single coil has two poles, north and south. Thus a three-coil alternator would usually be referred to as a six-pole alternator. Figure 4-5 shows a six-pole alternator and its output waveform. Notice that the output voltages of the six-pole alternator are not in phase with each other. They are respectively $60°$ out of phase, as you would expect from the way the coils are mounted. This alternator, having three separate output voltages out of phase with each other, is called a *three-phase* alternator. An alternator with only one armature coil is a *single-phase* alternator.

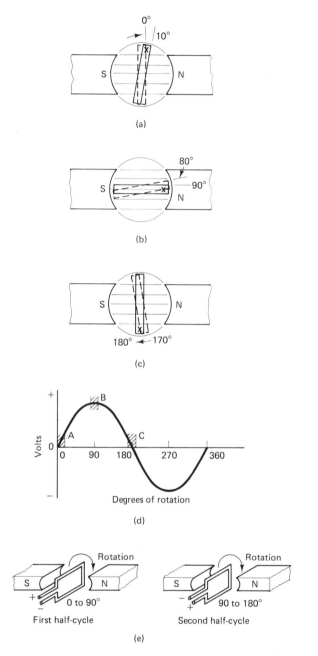

Figure 4-4 Output waveform of ac generator: (a) 0 to 10°; (b) 80 to 90°; (c) 170 to 180°; (d) graph of output voltage versus degrees of rotation; (e) polarity inversion.

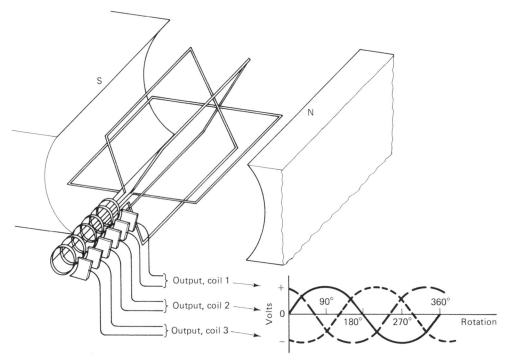

Figure 4-5   Alternator having three armature coils.

THREE-PHASE POWER SYSTEMS

Most alternators in common use today are of the three-phase variety, including the large ones used by commercial power generating companies. Most
industrial plants are fed by three-phase power. To avoid having to use six
wires to feed these plants, either of two power transmission systems can be
used, as shown in Fig. 4-6. The Y (sometimes called "wye") system uses
three *phase conductors* to carry the power; usually there is a fourth wire
called the *neutral*, which is often grounded to the earth. The delta system,
named for its resemblance to the Greek capital letter Δ, uses three phase
conductors only. There is no neutral. Obviously, a power company could
have a 25% saving in wire by using only delta systems. However, there is a
reason that this is not practical. Figure 4-7 shows a wiring diagram for the
ABC Widget Company's factory. Notice that the three-phase power is not
only fed to three-phase equipment, but is split up and fed to single-phase
equipment also. With a delta system, the current from each phase conductor
must equal the *phasor sum* of the current in the other two conductors. This
is another way of saying that in a single half-cycle of current, all the elec-

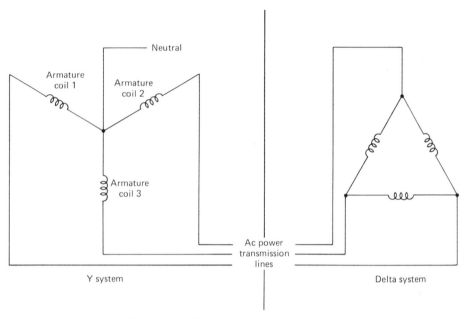

Figure 4-6   Three-phase power systems.

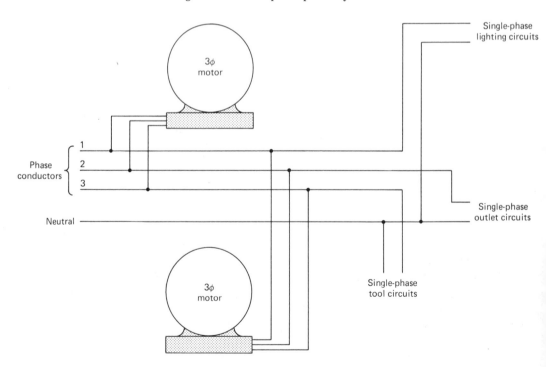

Figure 4-7   Three-phase factory wiring system.

trons entering the load (motors, lights, etc.) from conductor 1 must return to the alternator through conductors 2 and 3. There cannot be any leftovers. Hoarding electrons is not allowed. If only three-phase loads were used, this would be an easy condition to fulfill, because each phase of the load would automatically draw the same current as each other phase. But when single-phase loads are attached, there is no way to require Ed(na), the secretary, to turn off a typewriter (ϕ2 outlet circuit) just as Jo(e), the maintenance person, starts the electric drill (ϕ3 tool circuit). That is, there is no way to keep the loads always *balanced.* With a Y system, any unbalanced current drawn by one of the phase loads can return to the alternator via the neutral conductor. For this reason, power companies usually use delta systems only for transmitting power between their own generating plants and substations, then use Y systems for distributing power to individual consumers. Figure 4-8 shows this scheme.

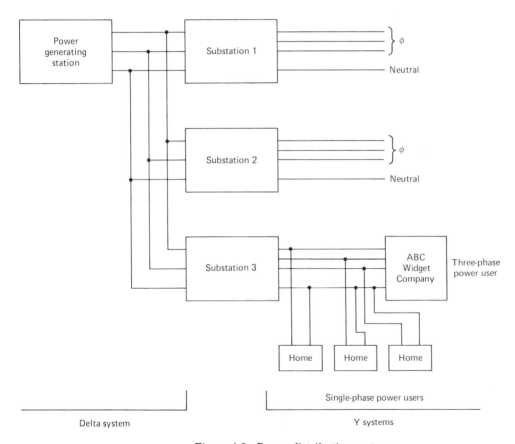

Figure 4-8   Power distribution systems.

ALTERNATORS

So far, all the alternators that we have discussed have been of the type in which the armature revolves. Since a voltage is induced because of *relative* motion between the armature and the field, it is equally possible to design alternators so that the armature is the stationary, or *stator* winding, and the field is the rotating, or *rotor* winding. In fact, by far most of the alternators in use are of the *revolving-field* variety. There is a very good reason for this. The current in a field winding is much less than the armature current. Consequently, by making the field winding revolve, only the smaller field current must pass through the slip rings and brushes. This makes for longer brush life, somewhat greater efficiency (electrical power output divided by mechanical power input), and less sparking at the sliding contacts. The high armature current passes through nonsliding permanent connections. Figure 4-9 is a simplified sketch of a two-pole revolving-field alternator. Figure 4-10 shows two typical automotive alternators of the revolving-field type.

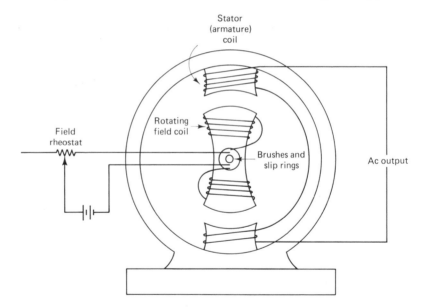

Figure 4-9   Two-pole revolving-field alternator.

Clearly, the best way of avoiding alternator problems caused by brush wear would be to do away with the brushes. Surprisingly, this can be done. Figure 4-11 is a simplified sketch of an *inductor alternator*, which has no sliding contacts. (Actual inductor alternators have windings on all four poles.) In this type of alternator, the relative motion between the magnetic field and the armature coils (called the secondary winding, in this case) is provided by the distortions of the magnetic field resulting from the moving

(a)

Figure 4-10 Automotive alternators: (a) Chrysler alternator (courtesy of Chrysler Corp.); (b) Delco alternator (courtesy of Delco-Remy Division of General Motors).

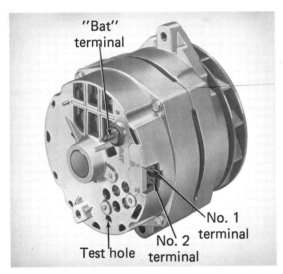

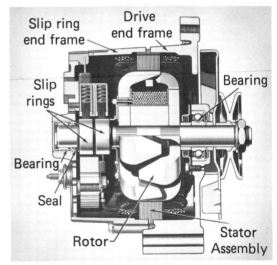

(b)

teeth of the soft-iron rotor. Notice in the figure that the field becomes alternately stronger and weaker, so that lines of force cut across the secondary winding, inducing ac into these windings. However, the number of lines of flux that are thus made to cross the armature coils is smaller than the number that would cross the coils in a revolving-field alternator. Therefore, in order to generate a sizable voltage, the relative motion must be faster. For this reason, inductor alternators are normally used to produce high-frequency ac.

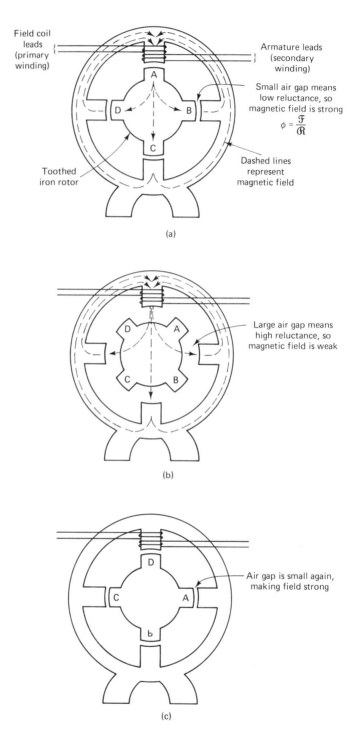

Figure 4-11 Inductor alternator: (a) $0°$ rotation; (b) $45°$ rotation; (c) $90°$ rotation.

RATINGS OF ALTERNATORS

Users of alternators must have a way of determining which alternator fits their specific needs. This can be determined from the manufacturer's ratings. These include:

1. The range of rpm at which the alternator is intended to operate
2. The nominal rpm
3. The voltage output obtainable at the nominal rpm for:
   a. No load
   b. Nominal load
   c. Some specified overload (often 150%)
4. The current rating
5. The number of poles
6. The number of phases
7. The field coil current and voltage requirement

(Often the voltage and current ratings are multiplied to give a rating in kilo-volt-amperes, or kVA.) The nominal load is whatever resistance is required to draw the rated current from the alternator. A 150% overload is whatever resistance is required to draw 150% of the rated current. For example, a 150-V 10-A alternator might be rated to produce 160 V at 3600 rpm, with no load. With the nominal load connected (150 V/10 A = 15 Ω), it would produce 150 V. With a 150% overload (15 A), it might produce 145 V. The overload specification is given to show how the alternator will perform if it is briefly connected to too low a load resistance. Continued operation in an overloaded condition will draw more current than the armature is designed to handle; overheating would be likely, possibly resulting in permanent damage. Maximum rpm could be 4500. Field current might be about 1 A at 12 V. Of course, less field current could be used to give a lower output voltage at 3600 rpm. Normally, no more than the rated field current would be used because of the possibilities of overheating and magnetic saturation. Knowing the frequency of the output voltage is also important. However, this can be found from the rpm and the number of poles per phase:

$$f_{\text{out}} = \frac{\text{rpm}}{60} \times \frac{\text{number of poles per phase}}{2}$$

Thus our alternator turning at 3600 rpm would have an output frequency of 60 Hz if it happens to be a two-pole single-phase alternator. A four-pole, two-phase or six-pole, three-phase alternator would still have two poles per phase, so the output frequency for each phase would still be 60 Hz.

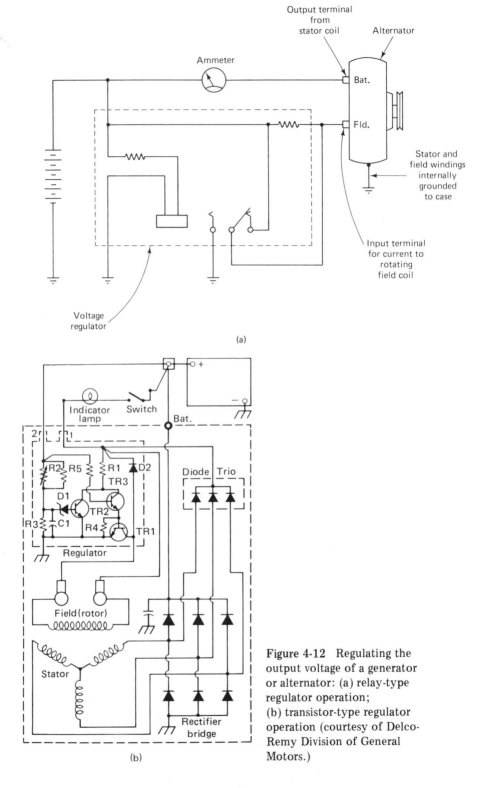

Figure 4-12 Regulating the output voltage of a generator or alternator: (a) relay-type regulator operation; (b) transistor-type regulator operation (courtesy of Delco-Remy Division of General Motors.)

FIELD EXCITATION

The simplest way of deriving field current for an alternator is from a battery. The battery is kept charged by the alternator's rectified, regulated output voltage. This method is used for an automobile alternator. The field voltage can be regulated either by a relay and resistor, or by an electronic voltage regulator. Both types of regulator are shown in Fig. 4-12. Notice that the control voltages for the relay and the electronic regulator are both taken from the alternator's output voltage. In this way, the output of the alternator is kept at its proper voltage, in spite of variations in rpm.

The regulators shown in Fig. 4-12 operate as follows:

1. *Relay-type regulator:* When the battery voltage is low, the regulator coil does not have enough current to close the normally open contacts, so battery voltage is fed through the dropping resistor to the field. As the charging voltage rises, sufficient voltage is applied to the coil to close the contacts, grounding the field winding at both ends. The voltage then falls as the battery is drained, so the contacts open again. In actual use, the contacts vibrate open and closed up to about 200 times per second, keeping the *average* voltage output steady. The battery acts somewhat as a capacitor to smooth out the resulting output waveform.

2. *Transistor-type regulator:* The circuit made of TR2, TR3, and their associated components senses the battery voltage, automatically adjusting the control current fed to the base of TR1. TR1 acts as an automatically varying resistance between the field coil and ground. In this way, field current is adjusted to whatever value is required to keep the output voltage steady.

In some alternators, the dc field current is provided by a small dc generator whose armature is mounted on the same shaft as the alternator's field winding.

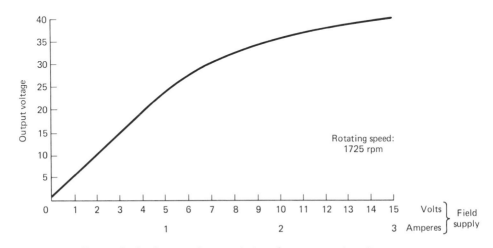

Figure 4-13 Output characteristics of an automotive alternator.

If the output voltage of an alternator is to be made variable, the field can be fed through a rheostat in order to make the field current adjustable. A graph of the output voltage of an automotive alternator versus the dc field current and voltage is given in Fig. 4-13.

PARALLELING ALTERNATORS

Many systems that are powered by alternators have two alternators so that one can be serviced while the other feeds uninterrupted power to the system. When the system is being switched from one alternator to the other, the output voltages, frequencies, and phase of the two alternators must be matched exactly. A circuit for determining when the outputs are matched is shown in Fig. 4-14. In use the circuit indicates an equal-voltage, equal-frequency, in-phase condition by no illumination of the indicator lamp. If the frequency, phase, or voltage are not equal, there will be times at which these differences produce a voltage difference across the indicator lamp. The voltages can be equalized by reference to the voltmeters. The frequency and phase are adjusted by varying the speed of the off-line alternator's engine. When the lamp alternates slowly between bright and dark, the frequency and phase are nearly equal. Then if the changeover is made during a dark period, no undesirable voltage spike will be produced on the output line.

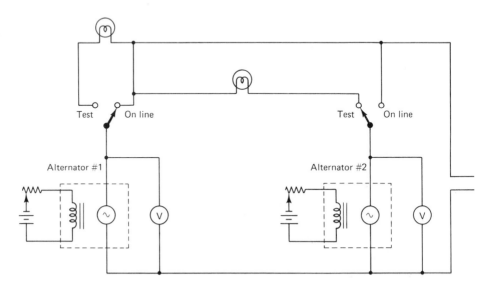

Figure 4-14 Test circuit for use when switching from one alternator to another.

LOAD CONSIDERATIONS

Since an alternator's armature is simply a strangely shaped iron-core coil, its impedance is mostly inductive. The resistance is quite small. Thus the current will lag the voltage by almost 90°. Depending on the type of load (that is, in terms of power factor), several things can happen to the output voltage as the output current varies. If the load is inductive, its current will also lag the voltage by about 90°, and the output voltage will drop as more output current is drawn. This drop is caused by the current flowing in the internal impedance of the armature. If the load is resistive, its impedance will be 90° different from the armature's impedance, reducing the total effect of the armature's inductance. Thus the drop in output voltage with increasing current will be less. If the load is capacitive, the load current will lead the voltage by 90°, so the load impedance will tend to cancel the effect of the armature inductance. The output voltage will therefore rise as output current increases until so much current is drawn that the armature resistance drops a significant amount of voltage. Then it will begin to fall. A graph of these three situations is shown in Fig. 4-15. An example of an inductive load would be a lightly loaded electric motor; a resistive load, a heating circuit; and a capacitive load, a rectifier feeding a large electroplating operation.

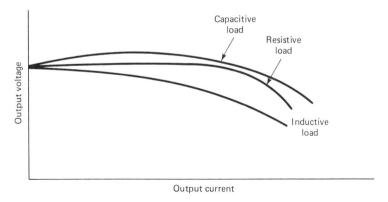

Figure 4-15   Variation of alternator's output voltage with load.

DC GENERATORS: BASIC OPERATION

Not all electric generators produce ac. There are many applications in which a dc generator is used. For many years, the battery-charging device in automobiles was a dc generator. Since high-current silicon rectifier diodes became available, alternators with built-in rectifiers have largely replaced generators in this application. But there are still applications for dc generators.

Going back to the basic generator—a flat armature coil rotating in a magnetic field—we remember that after each 180° of rotation, the polarity of the leads changes. By reversing the connections to the armature each 180°, we can keep the output polarity constant. This automatic reversal is accomplished by *commutators*, which are nothing more than segmented slip rings. Electrically, a commutator functions as a mechanical rectifier. As shown in Fig. 4-16, the output from the brushes is a pulsating direct current. In other words, although the polarity of the output voltage never changes, the amplitude changes from zero to maximum and back to zero twice per revolution. For most applications, this much pulsation in voltage is undesirable. The pulsation can be reduced by the use of additional armature windings, each connected to its own commutator segments, as shown in Fig. 4-17. In this two-coil—or four-pole—armature system, a winding is connected to the load only when it is cutting the maximum amount of flux. Thus the output voltage is the combined maxima of the individual coils' outputs. The output voltage has a higher average level, and never drops to zero. Any number of armature windings can be used; the more windings, the smoother the output voltage.

The actual winding pattern for an armature having a four-segment commutator is shown in Fig. 4-17b. In operation, when the brushes are contacting segments 1 and 3, the motor acts as if it had two armature coils connected in parallel: coil A + B and coil C + D. Segments 2 and 4 are not connected to the brushes at this time, so they simply serve as junction

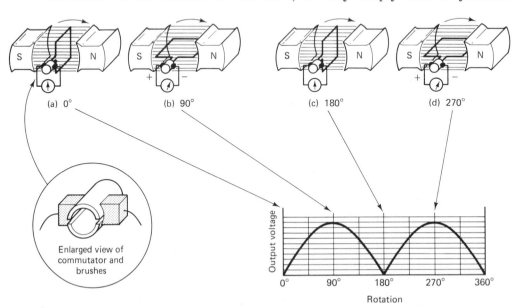

Figure 4-16 Operation of commutator and brushes.

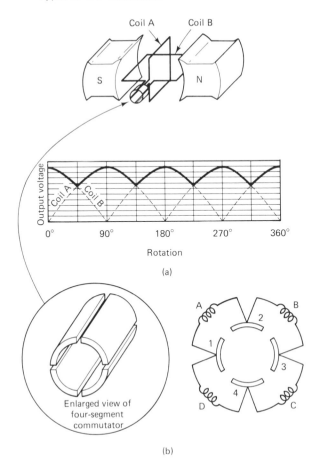

Figure 4-17 Output from four-pole generator: (a) theory; (b) schematic of the way an armature for a four-segment commutator is actually wound.

points. The net effect is the same as if there were one armature coil. When the brushes are contacting segments 2 and 4, the coil connection is changed: coil A + D is now paralleled by coil B + C. Thus the rotation of the armature changes the configuration in which the coils are connected, with the final result being the same as that obtained in part (a) of the figure. The advantage of this style of winding is that it makes the construction of an armature much simpler than if the configuration shown in Fig. 4-17a were used.

TYPES OF DC GENERATORS

Dc generators are grouped into several different types, depending on how the magnetic field is provided. The simplest type uses a permanent magnet and is called a *PM* generator. PM generators have several advantages:

1. For a given output capacity, a PM generator is smaller and lighter than its electromagnetic-field counterparts.

2. The output voltage is proportional to shaft rpm.

3. They are the most efficient type of generator, since no power is used to generate the magnetic field.

However, PM generators have one serious disadvantage. When a generator is producing a large output current, the electromagnetic effect of that current flowing in the armature can be quite large. The large magnetic fields generated in this way can demagnetize the field magnets, making the generator useless. This combination of characteristics makes the PM generator useful primarily in instrument applications. For example, PM generators are often used in tachometers. They convert a rotational speed to an output voltage, so that the speed of a machine can be measured by a properly calibrated voltmeter. In such an application, very little current is drawn, so there is no danger of demagnetizing the field magnets.

Generators whose magnetic fields are provided by an electromagnet fall into two categories. If the field windings are fed from an external current source, the generator is said to be *externally excited*. Externally excited field windings can be fed from a battery/rheostat circuit, or from a more complex circuit that provides automatic voltage regulation, as we discussed in connection with alternator field supplies earlier in this chapter.

Most dc generators are of the *self-excited* variety. There are several ways in which the field winding can be connected to obtain its current from the armature. Generators whose field windings are in series with the armature are called *series-wound* generators. If the field is connected in parallel with the armature, the generator is of the *shunt-wound* variety. Sometimes the field winding is made up of two electrically separate coils, and one is connected in series with the armature; the other, in parallel. Such generators are referred to as *compound-wound* (see Fig. 4-18). As you would expect, these different varieties of generators have different characteristics. Before we discuss these characteristics, we need to introduce a new concept.

Whenever a current is drawn from a voltage source, the output voltage of that source is likely to vary somewhat. Any voltage source, whether it is a battery, a generator, or whatever, has a maximum rated current output, referred to as the *full-load* current. The voltage at the terminals of the source when no current is being supplied (i.e., no load connected) is called the *no-load* voltage. Similarly, the voltage that appears at the terminals when the full-load current is being drawn is called the full-load voltage. The amount that the voltage varies as the load current changes is called the *voltage regulation*. It is usually expressed as a percentage:

$$\% \text{ voltage regulation} = \frac{V_{\text{no-load}} - V_{\text{full-load}}}{V_{\text{full-load}}} \times 100\%$$

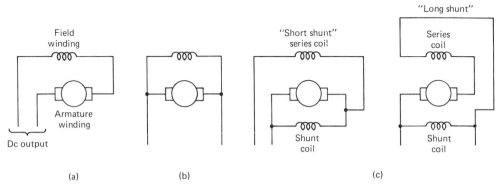

Figure 4-18 Varieties of electromagnetic-field generators: (a) series-wound; (b) shunt-wound; (c) compound-wound.

The series-wound generator has a voltage-regulation characteristic that may seem strange at first: The more current drawn by the load, the higher the output voltage. A moment's thought will show why this is so. With no current drawn, the field coils can produce no magnetic field, so the generator's output is essentially zero. (Actually, there is always a small amount of residual magnetism in the iron core of the field coils, so a small output voltage is generated even with no current in the coils.) As the output current increases, the field current increases, since the current is the same throughout a series circuit. Thus more output current makes for a stronger magnetic field. The stronger field, in turn, produces more output voltage. This process is limited by magnetic saturation: At some value of current, the iron core cannot support any stronger field. (Remember the hysteresis curve in Chapter 2?) Increases in current beyond this level cannot produce corresponding increases in output voltage. A typical graph of output current versus output voltage for a series generator is shown in Fig. 4-19a. This curve represents *negative* voltage regulation, as you can see from the graph and the equation above.

The field current of an ideal shunt-wound generator does not depend on the output current. This is true because the current in each branch of a parallel circuit is independent of the current in any other branch. In a real

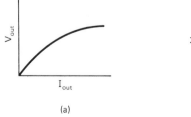

(a)

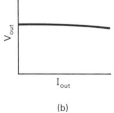

(b)

Figure 4-19 Output characteristics of series- and shunt-wound generators: (a) series-wound; (b) shunt-wound.

shunt-wound generator, the output voltage drops slightly at high output currents because of the internal resistance of the armature windings. This slight drop causes a slight decrease in field current as load current increases. Figure 4-19b shows an output graph for a shunt-wound generator.

Compound-wound generators offer an infinite variety of possible output characteristics. This is because the wire gauge and number of turns of the series coil and shunt coil can be varied independently. Also, it is possible to connect the two coils so that their magnetic fields aid (*cumulative compound*) or oppose (*differential compound*). The designer's choices in these areas determine whether the generator is *overcompounded, flat-compounded,* or *undercompounded.* As shown in Fig. 4-20, an overcompounded generator is one that acts more like a series-wound than a shunt-wound generator, and has a negative voltage regulation (V_{out} increases as I_{out} increases). A flat-compounded generator has essentially 0% (perfect) voltage regulation at or below the rated output current. An undercompounded generator has an output voltage that decreases slightly as the load current increases, much like a shunt-wound generator. All three of these types are of the cumulative-compound variety. A differential-compound generator has very poor voltage regulation and is used only for applications in which automatic limiting of the load current is needed.

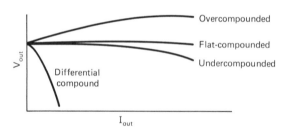

Figure 4-20 Output characteristics of compound-wound generators.

ARMATURE REACTION

It was mentioned during the discussion of PM generators that the magnetic field resulting from armature current must be taken into consideration by generator designers. This is also true for electromagnetic field generators, but for a different reason. For any armature coil, there is a certain position, called the *neutral plane*, at which the induced voltage in the coil is zero. In a two-pole generator, this should be the point at which the brushes briefly contact both commutator segments as they reverse the connection to those segments. (This process is called *commutation.*) In a four- or more pole generator, the commutation should occur at the point at which the induced voltages on the two adjacent segments are equal. If the coils are not passing through the neutral plane at the time of commutation, the different voltages on the two segments will result in spikes of high current through the brushes, normally accompanied by sparking. This sparking is undesirable from several standpoints, as we will discuss later. The important point for now is to understand the reason for wanting the commutation to occur when the armature coils are oriented in the neutral plane. With no current being drawn from the armature, this presents no problem to the designer. However, when there is current in the armature, the resulting magnetic field set up by the ar-

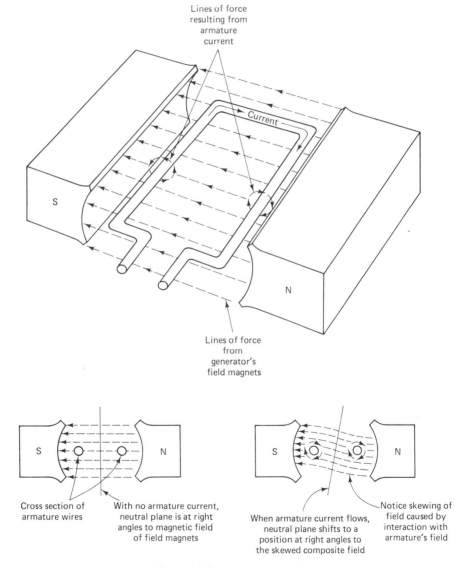

Figure 4-21 Armature reaction.

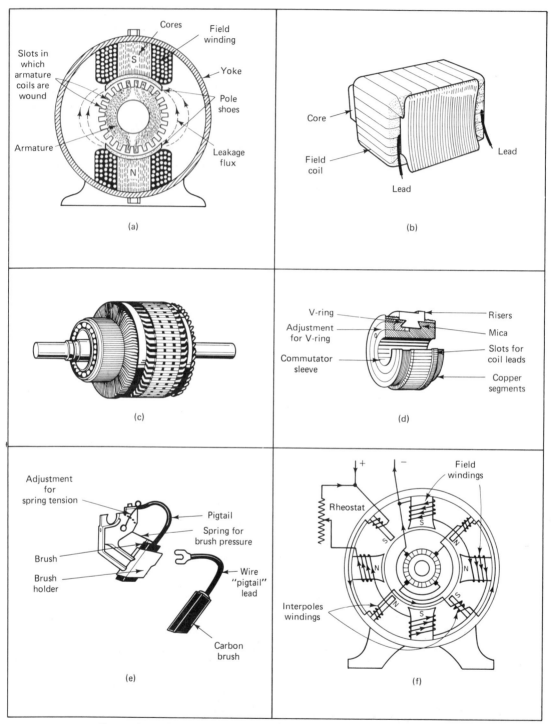

Figure 4-22 Parts of a dc generator: (a) magnetic circuit of a two-pole generator; (b) field coil on pole piece; (c) a complete armature; (d) commutator construction; (e) typical pigtail brush and holder; (f) schematic wiring diagram of
shunt generator. (Courtesy of the Bureau of Naval Personnel.)

mature distorts the field produced by the generator's field coils, causing the neutral plane to shift. This process, called *armature reaction*, is illustrated in Fig. 4-21. It is not practical to shift the brush positions to keep up with the moving neutral plane as the generator's load current changes. Therefore, an electromagnetic method was devised to correct for armature reaction. This method involves the use of separate electromagnets, called *interpoles* or *commutating poles*, to cancel out the effect of armature reaction. Regardless of the type of generator, the interpole winding is always connected in series with the armature. Thus when there is no current drawn from the armature, there is also no current in the interpole windings, and the neutral plane is in its normal position. As armature current increases, the current in the interpole windings also increases, so that the magnetic field generated by the armature is exactly canceled by the field from the interpole windings. Figure 4-22 shows drawings of the various parts of a shunt-wound dc generator, including the interpole windings. A series-wound generator would look virtually the same, except that the internal connections diagrammed in part (f) of the figure would be different.

MAINTENANCE

Maintenance and repair of alternators and generators is essentially the same as for motors, and will be discussed fully in a later chapter. There is one repair item, however, that is sometimes required on alternators and generators that does not apply to motors. This is the testing and replacement of rectifiers and voltage regulators that are associated with the alternator or generator. Most often, these items are built into automotive alternators only. For other types of alternators, they are likely to be mounted separately.

If the output voltage of a generator or alternator is incorrect, the voltage regulator may be at fault. If the generator voltage is too high, the regulator circuit is almost certainly defective. If the output voltage is too low, further testing is required.

There are two tests for voltage regulators. The first is an operational test to determine whether the regulator is passing field current. Simply measuring the field terminals on the generator or alternator with a voltmeter will tell you whether a field voltage is being applied. If there is no field voltage, the regulator is probably defective. If the field voltage is present, the next test is to temporarily disconnect the generator from the regulator and to replace the regulator with a rheostat. This should permit you to adjust the generator output voltage from too low a value to too high a value. If the output voltage still cannot be made high enough, the problem is in another part of the field supply (such as a ballast or dropping resistor) or in the generator itself. If the voltage can be made too high, the regulator needs to be adjusted or replaced. Mechanical regulators can be cleaned and adjusted as discussed in Chapter 3, via the return spring tension adjustment. Some electronic regulators can be adjusted by means of a rheostat or potentiometer. Others are nonadjustable and must be repaired or replaced if they are not operating within their specifications. Many of these units are encapsulated in silicon rubber or epoxy and cannot be repaired. Some, however, can be repaired by a competent electronics technician. Discussion of repair techniques for electronic voltage regulators is beyond the scope of our present study.

If the field supply circuit includes a rheostat rather than a voltage regulator, it may be defective. Testing a rheostat can be done by means of an ohmmeter. One probe is connected to each terminal of the rheostat. The ohmmeter range is selected to give about a half-scale reading at the rheostat's maximum resistance. If the resistance is not marked, some trial and error may be required to find the proper range setting. As the shaft is rotated smoothly from minimum to maximum, the ohmmeter indication should increase smoothly from nearly zero ohms to a value equal to the resistance of the rheostat. If the meter indication is erratic, or if there are points at which the resistance becomes very high ("dead spots"), the rheostat should be cleaned with a silicon-lubricated spray-type contact cleaner. If the problem does not clear up, the rheostat must be replaced. It is important that a replacement unit of equal resistance and equal or greater power rating be used. Only wirewound rheostats should be used for regulating field coil current, since other types do not have sufficient current-handling capability. Occasionally, it may be necessary to replace a rheostat for which neither the resistance nor the power rating is specified. If the rheostat has dead spots but is not burned open, the resistance can be determined with an ohmmeter. If it is open, the resistance can be found by removing the back cover of the rheostat so that the resistance wire is visible. Usually, the point at which the open occurred is visible as a black spot or a melted point in the wire. Frequently, this place is near the fixed terminal of the unit. The ohmmeter can then be used to measure the resistance of the remaining wire. By estimating the fraction of the total wire length that was measured, the original resistance can be determined. For instance, if the wire opened one-third of the distance from the fixed terminal to the far end, and the resistance from the fixed terminal is 31 Ω, the original resistance was probably 100 Ω. (31 $\Omega \times 3 = 93 \Omega$, so the nearest standard value is 100 Ω). The power rating can usually be approximated quite closely by matching the rheostat with a new one of similar appearance and the same or larger physical size.

If the regulator is determined to be operating correctly but the output voltage is too low, the brushes should be checked. Worn, pitted, cracked, or oil-soaked brushes must be replaced.

If all else fails, the rectifiers and windings should be tested. Methods of testing windings of motors, alternators, and generators are discussed in Chapter 8. The subject of testing rectifiers requires a brief introduction.

As we mentioned earlier, automotive alternators are three-phase (or six-pole) types. Their output is converted from ac to dc by means of rectifiers

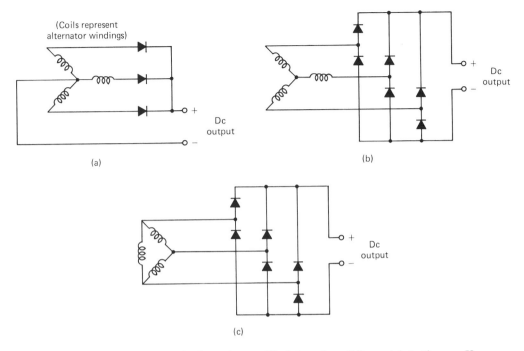

Figure 4-23 Circuits of alternators with integral rectifiers: (a) half-wave Y; (b) full-wave Y; (c) full-wave delta.

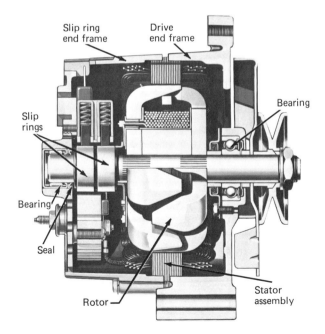

Figure 4-24 Cross-section of a Delco alternator. (Courtesy of Delco-Remy Division of General Motors.)

built into the alternators themselves. The circuit of such a system is shown in Fig. 4-23. Testing these rectifiers is done by removing them from the circuit so that at least one terminal of each rectifier is not connected to anything. Then an ohmmeter is used, set on the R × 1000 scale, to measure the resistance between the terminals of each of the six rectifiers. For each rectifier, the ohmmeter should indicate a low resistance with the probes connected with one polarity, and an open circuit (infinite resistance) with the probes connected in the opposite polarity. A low resistance in both directions indicates a shorted rectifier. A high resistance in both directions indicates an open rectifier. Either must be replaced. Figure 4-24 shows a disassembled Delco alternator, and Fig. 4-25 shows the testing method for a Delco integral rectifier assembly.

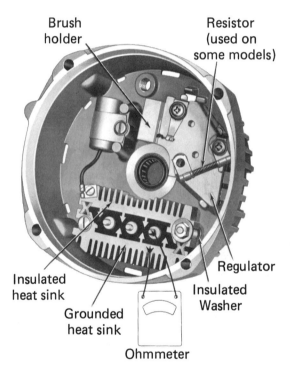

Figure 4-25   Testing method for rectifiers. (Courtesy of Delco-Remy Division of General Motors.)

SUMMARY

1. Generators convert mechanical energy to electrical energy. They consist of a field (permanent magnet or electromagnet) and an armature, consisting of coils in which the output current is generated.

2. Connections to rotating armature coils can be made by brushes and either commutators or slip rings.

3. Ac generators are called alternators and usually have three-coil (six-pole) armatures. Such alternators produce three-phase outputs, which can be connected in either delta or wye configurations. Alternators are usually built so that the field revolves and the armature is stationary. Inductor alternators have no connections to the rotor. Alternators are rated in rpm, voltage and current or kVA, number of poles, number of phases, and field coil requirements. Output frequency can be obtained from the rpm and number of poles per phase. Alternator output voltage is regulated by regulating the field current. Alternators used in multiple-alternator systems must be operating at equal voltages and frequencies with their outputs in phase when the load is switched from one alternator to the other. An alternator's voltage regulation depends on the output current and the power factor of the load.

4. Dc generators are like alternators with commutators instead of slip rings. The more poles a generator has, the smoother is its dc output. Types of generators include PM, series-wound, shunt-wound, and the several types of compound-wound. Each has its own characteristics, advantages, and disadvantages. Armature reaction in dc generators is compensated for by interpole windings.

5. In addition to the maintenance items encountered in motor work, generators also occasionally require adjustment, repair, or replacement of voltage regulators, field rheostats, and rectifiers.

QUESTIONS

1. (Fill in the blanks). A generator converts _____ energy into _____ energy.
2. What are two ways of providing a generator's field?
3. What is an armature?
4. What is the purpose of:
   (a) Slip rings?
   (b) Brushes?
   (c) Commutators?
5. Why does a generator without a commutator produce ac?
6. How many windings are there in a four-pole alternator? How many different phases of ac would it produce?
7. What is the difference between Δ and Y power systems? Which type is inherently balanced? Why?
8. Why are revolving-field alternators commonly used?
9. What type of alternator has no brushes? How does it work? What are alternators of this type primarily used for?

10. What is the nominal load for a 117-V 40-A alternator?

11. True or false: The performance of an alternator can be improved by using a higher-than-rated field current.

12. True or false: Using too low a field voltage will damage a generator.

13. Explain the operation of a relay-type voltage regulator used with an alternator.

14. In what two ways can the output voltage of a generator be varied?

15. Why is a special circuit required to switch from one operating alternator to another?

5

DC MOTORS

The basic operating principle of all electric motors is the same: the force exerted on a current-carrying conductor that is in a magnetic field. However, the hardware used to harness this force is different for a dc motor than for an ac motor. In this chapter we look at dc motors, and much of our study will sound familiar. This is because the construction of dc motors is quite similar to that of the dc generators that we studied in Chapter 4. In fact, most dc motors can be used as generators, and vice versa.

DC MOTOR ACTION

Figure 5-1 illustrates the motor principle acting on a linear conductor. The resulting force pushes down on the conductor. If the conductor were formed into a loop, the force would tend to cause the loop to rotate, as shown in Fig. 5-2. This rotational force, or *torque*, will continue until the coil has rotated 90° and is perpendicular to the magnetic lines of force (Fig. 5-3). This is called the neutral plane, just as for a dc generator. When the coil is in that position, the magnetic force on the two halves of the coil will cancel out, but the coil will continue to rotate a bit further, because of its momentum. However, as soon as it passes the 90° point, the coil will experience a force pushing it back in the direction from which it came (Fig. 5-4). The result is that the coil is magnetically held at the neutral position. There is not much market for motors that rotate somewhat less than a half-turn at most and then stop. Therefore, a means was devised to reverse the battery connections to

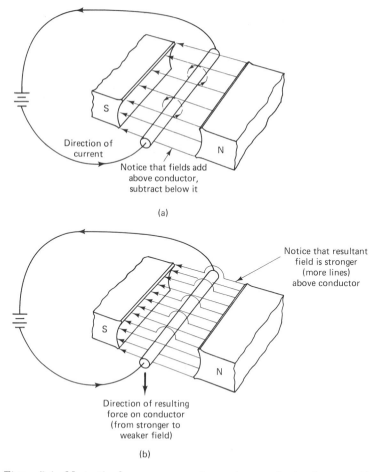

Direction of
current

Notice that fields add
above conductor,
subtract below it

(a)

Notice that resultant
field is stronger
(more lines)
above conductor

Direction of resulting
force on conductor
(from stronger to
weaker field)

(b)

Figure 5-1 Magnetic force on current-carrying conductor in magnetic
field: (a) magnetic fields of magnets and conductor shown separately;
(b) resultant net field.

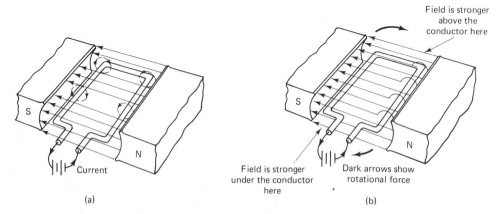

Field is stronger
above the
conductor here

Current

Field is stronger
under the conductor
here

Dark arrows show
rotational force

(a)   (b)

Figure 5-2 Rotational force on current-carrying coil in magnetic field: (a) magnetic fields of magnets and conductor shown separately; (b) resultant net field.

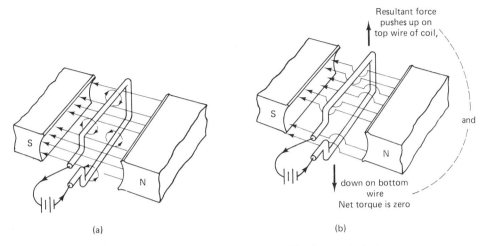

Figure 5-3   Force on a coil positioned in the neutral plane.

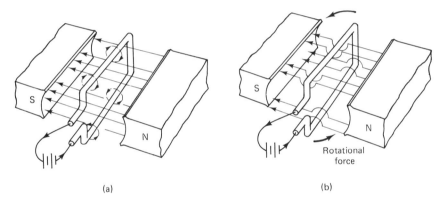

Figure 5-4   Force on a coil after it passes the neutral position: (a) fields shown separately; (b) resultant net field.

the armature coil each time the coil rotates past the neutral plane. This is our old friend, the commutator. By reversing the direction of the current in the armature, we also reverse the direction of the torque. Thus the action of a dc motor can be summarized as follows:

1. Current is applied to the armature, setting up a magnetic field. This field interacts with the field from the motor's magnets to produce a torque. The armature begins to rotate.
2. When the armature coil becomes perpendicular to the magnets' field, its interaction with that field produces no net torque, but the armature's momentum carries it on past the neutral plane.

3. As the armature passes the neutral plane, the commutator reverses the direction of current in the armature coils, providing a new torque to keep the armature turning in the same direction.

With this type of motor, the torque varies from zero to its maximum value twice in each revolution. This variation in torque is undesirable for two reasons. First, it can result in excessive vibration of the motor and whatever equipment it drives. Second, if the motor should stop with the armature coil positioned in the neutral plane, there would be no torque to start it when power was reapplied. By using additional armature windings, each connected to its own commutator segments, both problems are solved. As shown in Fig. 5-5, the more coils an armature has, the smoother the torque output. And with two or more coils, the torque never drops to zero. For these reasons, the lowest number of armature coils that will be found in a commercial electric motor is two. (Two coils correspond to four poles, you remember.) Where a very smooth torque characteristic is essential, upward of 100 poles may be used. An example of such an application would be dc servo-controlled turntable motors as used in stereo music systems.

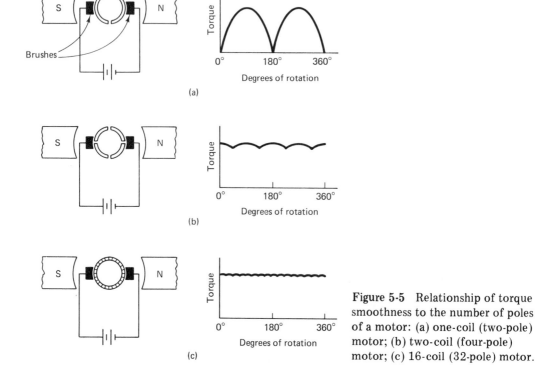

Figure 5-5 Relationship of torque smoothness to the number of poles of a motor: (a) one-coil (two-pole) motor; (b) two-coil (four-pole) motor; (c) 16-coil (32-pole) motor.

COUNTERVOLTAGE

If we apply dc to a motor through an ammeter, we will see a curious thing. When the circuit is first closed, a large current will flow, in accord with Ohm's law. But as the motor speeds up, the current will drop to a small fraction of the original current. This is illustrated in Fig. 5-6. The reason for this is the generator principle. You remember that whenever a conductor moves through magnetic lines of flux, a voltage is induced in the conductor. The conductor here is the armature, and the motor's field supplies the magnetic lines of flux. Lenz's law tells us that the induced voltage will be opposite in polarity to the applied voltage. Therefore, this voltage is called the motor's countervoltage. Whenever two voltage sources are connected in series, with polarities opposite, the resulting current will follow the polarity of the larger source and will be proportional to the magnitude difference between (i.e., the *algebraic sum of*) the two sources. The way this applies to our motor is shown in Fig. 5-7. At 0 rpm, the induced voltage is zero, so the current is given by $I = E/R = 6 \text{ V}/3 \text{ } \Omega = 2 \text{ A}$. At 1750 rpm, the induced voltage is 5.7 V, so

$$I = \frac{6 \text{ V} - 5.7 \text{ V}}{3 \text{ } \Omega} = 0.1 \text{ A}$$

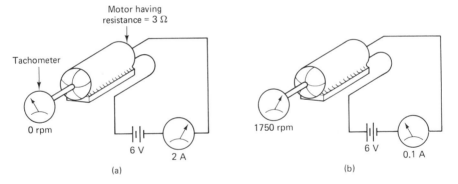

Figure 5-6 Variation of motor current with speed: (a) starting; (b) running.

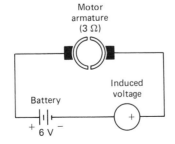

Figure 5-7 Effect of induced voltage in a motor armature.

Countervoltage plays a very important part in a motor's operation. When a motor is operating under a normal load, the speed is sufficient to generate a large countervoltage. Thus the motor current is limited to a safe value and overheating of the armature is avoided. When a motor is heavily loaded, the rotational speed is reduced and the countervoltage decreases. The motor then draws more current, creating stronger magnetic fields to increase the torque. If a motor is seriously overloaded, insufficient countervoltage will be induced to limit armature current to a safe value, and the motor may overheat and be destroyed.

ARMATURE REACTION

In a dc motor, just as in a dc generator, the current in the armature produces a magnetic field that interacts with the field of the stator. The resulting field has a neutral plane, that is, a plane at which the brushes can be located so that the countervoltages are equal on each commutator segment at the time that the brushes short them together. When the brushes are positioned in the neutral plane, brush sparking is at a minimum and brush life will be at a maximum. However, as the speed of the motor changes with variation in load, the armature current also changes (because of changes in countervoltage). Thus the strength of the armature's magnetic field changes, and the neutral plane shifts. To cancel the effect of this shift, interpole windings connected in series with the armature are used in some dc motors. The larger the motor, the more likely it is to have interpoles. Physically, these are exactly like the interpoles used in generators that were discussed in Chapter 4.

PM MOTORS

Having discussed the principles of dc motors in general, we will now look at several specific types. The first is the permanent-magnet, or PM, motor. This simplest type of dc motors uses a permanent magnet for its field. There are a number of advantages in doing so. Figure 5-8 compares a ceramic-magnet-field PM motor with a wound-field motor of the same horsepower. The total volume occupied by the PM motor is about half that of the wound-field motor. PM motors are correspondingly lighter in weight than comparable wound-field motors. In fact, the PM design gives the smallest and lightest motor available for a given horsepower.

Another advantage of PM designs is slightly greater efficiency. Since no power is consumed to provide the motor's field, a saving of about 10 to 15% is to be had.

A third advantage is relative freedom from armature reaction. The reason for this can be understood if the magnetic circuits of PM and wound-field motors are compared. In a wound-field motor, the magnetic lines of force generated by the armature current have a low-reluctance path (through

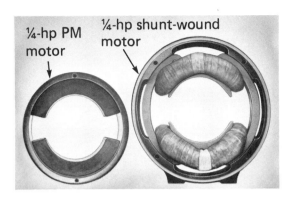

¼-hp PM motor

¼-hp shunt-wound motor

Figure 5-8 Size comparison—PM versus electromagnetic-field motors. (Courtesy of Bodine Electric Co.)

the iron field structure). The magnetic Ohm's law tells us that this results in a large flux density. In a PM motor, the armature's lines of force have a high-reluctance path through the ceramic magnets, which are a high-reluctance material. The armature's field then has a very small flux density, so it can only slightly distort the high-flux-density field produced by the magnets. Thus the neutral plane does not shift significantly.

For applications in which a motor must be reversible, PM motors are especially convenient. They can be reversed simply by reversing the polarity of the applied voltage. This can be done with the motor either at rest or in motion.

Sometimes, it is necessary for a motor to stop rotating very quickly after the power supply is disconnected. This can be arranged either by mechanical braking (friction), or electrical—often called *dynamic*—braking. Dynamic braking is accomplished in a PM motor by shorting the armature connections either directly or through a low-value resistor. This converts the motor into a PM generator and quickly uses up the rotational mechanical energy by converting it to electricity, then to heat. By this simple means, a PM motor can be braked very quickly without the use of mechanical linkages and brake shoes that can wear out.

A final advantage of PM motors is their very high starting torque. Unfortunately, this torque results from a high starting current (as much as 10 to 15 times the normal running current), which can be a *disadvantage*!

The only other serious disadvantage is the ability of permanent-magnet fields to be demagnetized by the very high armature currents that can result from stalling, or "locked rotor" operation. This problem is most significant at temperatures below 0°C.

One very important characteristic of any motor is the speed regulation, or the way in which speed varies as more or less torque is required by the load. This characteristic is best described by the speed/torque curve. Figure 5-9a shows how to interpret the information given by a speed/torque curve. Notice the "full-load speed" and the "rated torque." These characteristics are specified for each particular motor by the motor manufacturer.

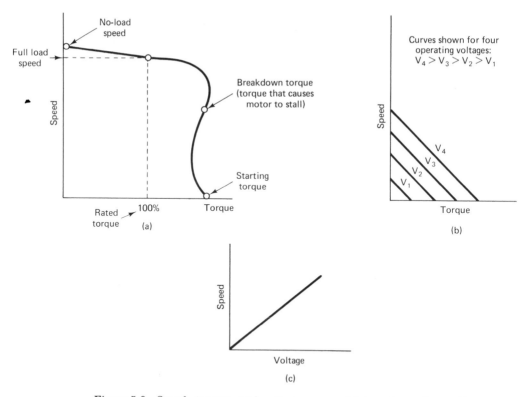

Figure 5-9 Speed, torque, and voltage curves: (a) how to read speed/
torque curves; (b) speed/torque curves; (c) speed/voltage curve.

Therefore, they will not be shown on the examples of speed/torque curves
that we will show in this book. Figure 5-9b shows speed/torque curves for a
typical PM motor. Curves are shown for several different values of applied
voltage. Figure 5-9c shows the effect of varying voltage, while the torque required to rotate the load remains constant. Notice several things:

1. The speed bears an inverse linear relationship to the amount of
    torque required by the load. Another way of saying this is that the
    torque available to drive a load increases as the heavier load slows
    down the motor. Notice that the starting torque and the breakdown
    torque are the same for this motor.

2. The speed is linearly related to the applied voltage. This makes
    speed control very simple; you just vary the supply voltage.

3. With no load, the motor speeds up only to a certain maximum rpm.
    As we will see, there are some motors that will "run away," or build
    up speed until their bearings or brushes are destroyed if they are
    operated without a load. All these characteristics can be either ad-
    vantages or disadvantages, depending on the application. Usually,

designers choose PM motors for applications at which the variation of speed with load can either be tolerated (as in weed cutters) or compensated for (as in dc servo-controlled tape decks—see Chapter 7).

A special application for PM motors that has appeared in the last few years is in low-cost portable tape players. In these applications, constant speed is provided by a centrifugally operated governor switch that disconnects power to the armature when the desired speed is reached. In operation, the switch opens and closes many times each second, providing "chopped" dc whose average voltage is just right to keep the motor turning at the desired speed. A governor-controlled PM motor is pictured in Fig. 5-10. Figure 5-11 shows a typical circuit for a PM motor to be used in battery- or

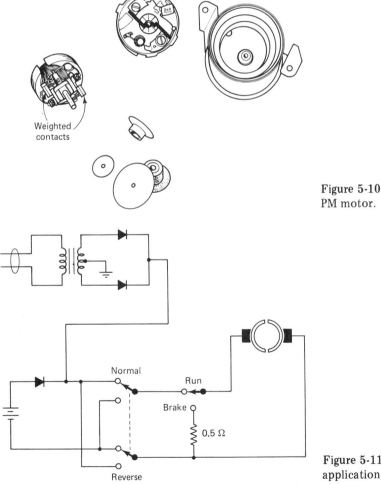

Figure 5-10   Governor-controlled PM motor.

Figure 5-11   Typical circuit for application of PM motor.

ac-line-powered operation, with reversibility and dynamic braking. Speed control could be provided by using a variable-voltage supply, or a rheostat in series with the armature.

SHUNT-WOUND MOTORS

A shunt-wound motor is one whose magnetic field is provided by an electromagnet connected in parallel (i.e., in shunt) with the armature. This connection may be inside the motor, or the four leads (two armature leads plus two field leads) may be brought out for connection externally. The primary advantage of a shunt-wound motor is its good speed regulation: Variations in torque required by the load do not make a great difference in speed unless the motor is overloaded. This is shown in the speed/torque curve in Fig. 5-12. The National Electrical Manufacturers' Association (NEMA) has agreed on four standard speeds for shunt-wound dc motors: 1140, 1725, 2500, and 3450 rpm.

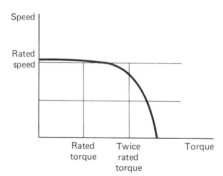

Figure 5-12   Speed/torque curve for shunt-wound motor.

Shunt-wound motors share with PM motors the characteristics of easy reversibility (at rest or in motion) and ease of dynamic braking. The only difference is that for shunt-wound motors, the field must be kept energized in the same polarity during these operations. This means that reversing the shunt-wound motor involves reversing the electrical connections to the armature only. Although it would be possible to reverse the connections to the field winding instead, this is usually not done, because of the large voltage spikes that can result from reversing the supply to such a high-inductance coil.

The starting torque of a shunt-wound motor is less than that of other types of dc motors of the same horsepower. This can be seen from the speed/torque curve; when the motor is heavily overloaded, it may not produce enough torque to keep turning and may stall. Such stalled operation can result in overheating of the armature windings. However, such overheating is not as serious a problem with shunt-wound motors as with other types of dc motors. This is because of the shunt-wound motor's low starting current, typically about three times the full-load running current.

The speed of a shunt-wound motor can be controlled in either of two ways. The most common method is to vary the armature supply voltage. The speed of the motor varies linearly with the armature supply voltage. The torque of the motor is not affected by this type of speed control. To avoid having to use a variable-voltage dc supply, the "shunted-armature" circuit of Fig. 5-13 can be used. The advantage of this circuit over a simple rheostat in series with the armature is better speed regulation in spite of varying loads. If only a rheostat were used, the following things would happen:

1. A heavy load would cause the motor to slow down.

2. Less countervoltage would be produced.

3. Armature current would increase as a result.

4. The increased armature current would result in greater voltage drop across the rheostat.

5. Less voltage would be available to run the armature, resulting in a further speed decrease.

The use of the circuit shown in Fig. 5-13 results in speed regulation being equally good at any speed.

Another speed control method that is sometimes used is to decrease the shunt-wound motor's field current. Although one might expect this to slow the motor down, actually it has the reverse effect. The reason is that, with a weaker field, the motor produces less countervoltage in the armature. Thus the armature will draw more current, causing the motor to speed up. There are three reasons that this method is seldom used. First, the increased armature current can cause overheating. Second, the speed regulation is poorer than with normal field current. Third, the motor's torque is decreased. These disadvantages limit the use of the *field weakening* method of speed control

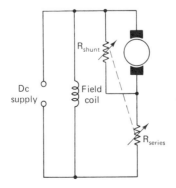

Figure 5-13 Shunted-armature speed control for shunt-wound motors. Dashed line indicates that both potentiometers are controlled by the same shaft.

to applications requiring a motor to turn a light load at a speed higher than the motor's rated speed.

One problem sometimes occurs in the operation of shunt-wound motors. If the field current should be interrupted at a time when the motor is turning but not loaded, the motor will increase its speed, possibly enough that bearings and/or brushes are overheated and destroyed. This uncontrolled speed increase is referred to as *runaway*. Runaway in a shunt-wound motor occurs because of a principle that we will explain in the next chapter when we discuss induction motors.

When dc motors are used to drive very heavy loads, the resulting large starting currents can cause problems. Even though the starting current for a shunt motor is considered low, it is still several times the running current. To limit starting current to a safe value, simple electromechanical systems called *motor starters* are used. The type of starter used with a shunt-wound motor is called a three-point starter. (It could more accurately be called a three-lead starter.) This type of starter is illustrated in Fig. 5-14. In operation, the switch is rotated slowly from position 1 (off) to position 7, with a brief pause at each position for the motor to attain maximum speed at that position. This connects progressively smaller resistances in series with the armature until, at position 7 (full speed), there is no series resistance connected. The electromagnet then latches the switch in position 7. Notice that if the

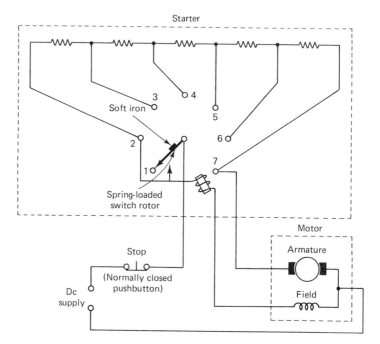

Figure 5-14 Three-point starter for shunt-wound motors.

field-circuit supply should open, the electromagnet will release the switch rotor, allowing it to return to the OFF position. This feature prevents runaway.

The combination of good speed regulation and fair starting torque make the shunt-wound motor suitable for applications in which the motor is not required to start against a heavy load. Examples are fans and instrumentation requiring a fairly constant-speed drive unit.

SERIES-WOUND MOTORS

By far the most common type of electric motor manufactured is the series-wound type. As its name indicates, this motor has its field coil connected in series with its armature. The advantages of series-wound motors are numerous. The most important are the size and weight—and consequently price. Since the field coil has a high current available (i.e., the full armature current), not many turns are required. Even though a somewhat heavier wire is used than for shunt-wound fields, the result is still a substantial saving in the amount of (expensive!) copper used. Also, the smaller field coils permit the use of a smaller laminated-iron field structure. The end result is that the series-wound motor provides the most horsepower per dollar of any type of motors. It is second only to the PM motor in horsepower per pound and horsepower per cubic inch. For instance, the modern electric drill would not be practical were it not for series-wound motors. With any other type of motor, the size and weight would make them of interest only to members of the NFL!

Because of the small number of turns (hence, low inductance) in the field coils, a series-wound motor can be used for either ac or dc. Since the polarity of both the field and the armature coils reverse simultaneously when ac is used, the resulting magnetic force is in the same direction. Because of its ac-or-dc operation, the series motor is often called a *universal* motor. (Actually, the inductance of the windings will give a series motor slightly less horsepower when operated from ac than it has when running on dc.) Series-type motors can be reversed by reversing the connection of the armature leads with respect to the field, as shown in Fig. 5-15. However, since armature reaction is a significant problem in series motors, and since interpoles are seldom used except on the largest ones, the brushes are located for an average neutral-plane position. If a series motor that is designed for one-

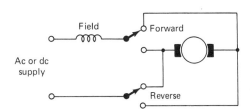

Figure 5-15   Reversing connection for series-wound motors.

directional operation is reconnected for reverse rotation, the resulting reversed armature reaction will cause serious brush sparking. The result will be poor torque and very short brush life. In short, series motors can be designed to be reversible, but should not be reversed unless they *are* so designed. Series motors are generally not adaptable to dynamic braking.

Many things can be learned from the speed-torque curve for serieswound motors, shown in Fig. 5-16. First, as speed is decreased by a heavy (high-torque) load, the motor supplies increasing amounts of torque to drive that load. The starting torque (at zero speed) is very high. This helps to prevent stalling and to allow the motor to start heavy loads. The starter motors used for cranking automotive engines are often series motors. As you would expect, the starting current is also high. Second, the speed regulation under varying load conditions is very poor. Third, when a series motor is operated with no load, extremely high speeds—possibly runaway—can result. These characteristics make series-wound motors especially useful for portable tools and appliances. Often, a series motor is connected to a built-in gear train to reduce the normally high shaft rpm and to provide even more torque. This has the added advantage that a gear train provides some loading and thus prevents runaway. Figure 5-17 is a photograph of an electric drill, disassembled to show the motor and gear train assembly.

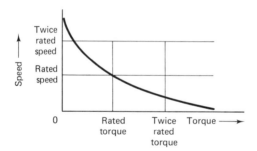

Figure 5-16 Speed/torque curve for series-wound motor.

The operating speed of a series-wound motor is normally 7000 rpm or more. Speeds up to 25,000 rpm are not uncommon in routers, miniature grinders, and orbital finishing sanders. These high speeds provide excellent results in these applications, but also result in considerably shorter life for brushes and bearings than for other types of motors. For this reason, series motors are usually given an "intermittent duty" rating. All other types of motors are usually designed for continuous duty. What this really means is that series motors require more frequent maintenance than the other types.

Controlling the speed of a series-wound motor is done very simply: The supply voltage is varied. The relationship of no-load speed to supply voltage is very similar to that for PM motors. Where a variable-voltage supply is not available, a rheostat can be used as shown in Fig. 5-18a or b. For series motors operated from ac supplies, variable transformers or electronic speed

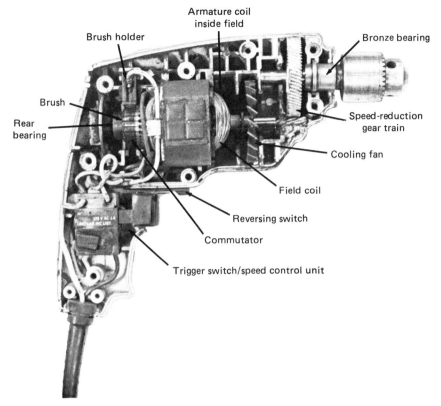

Figure 5-17 Inside an electric drill.

controls are most often used. Electronic speed controls are discussed in Chapter 6.

Series-wound motors are usually not made in large enough sizes for their high starting current to be a problem. However, there are a few applications in which a device is used to limit starting current. These devices come

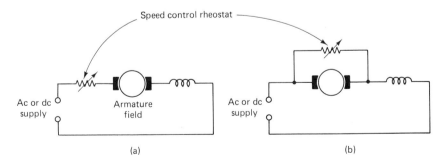

Figure 5-18 Speed control of series-wound motors.

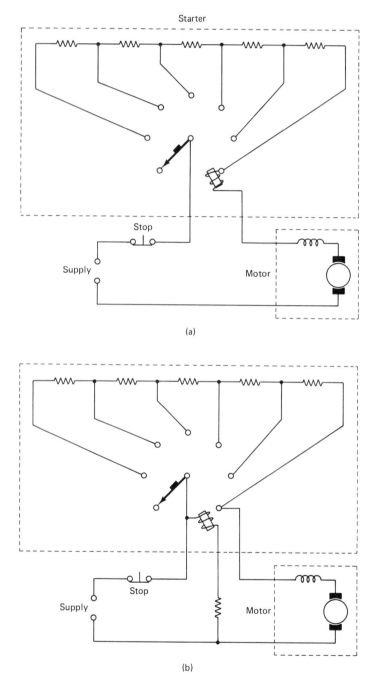

Figure 5-19 Starters for series-wound motors: (a) two-point; (b) three-point.

in two varieties (Fig. 5-19), called two- and three-point starters, respectively. Notice that the two-point starter will release the electromagnet if the current drawn by the motor becomes too small. Since a small current represents a light load, this feature disconnects the motor if it is unloaded, preventing runaway. Three-point starters do not offer runaway protection.

COMPOUND-WOUND MOTORS

Just as compound-wound generators can provide a combination or hybrid of the characteristics of series-wound and shunt-wound generators, compound-wound motors can provide characteristics that are a mixture of series-wound and shunt-wound motors' characteristics. To review briefly, a compound-wound machine is one in which the field coil is split into a segment connected in series with the armature and a segment in parallel with the armature. These are called the *series* and *shunt fields*, respectively. They can be connected either short-shunt or long-shunt (Fig. 4-18). The coils' magnetic fields can either aid (cumulative compound) or oppose each other (differential compound).

The speed/torque characteristics of compound-wound motors are shown in Fig. 5-20. Clearly, these characteristics provide for very different applications. For a cumulative-compound motor, the speed depends on the sum of the two fields. This in turn depends partially on the armature current, since the series coil's current *is* the armature current. When the motor is lightly loaded and the resulting countervoltage is large, the armature current is small. The speed is then determined mainly by the shunt field, and the motor "idles" almost like a shunt-wound motor. This means that there is no tendency to run away. When the motor is more heavily loaded, the series field comes into action and the motor begins to offer more and more torque as speed is decreased by the load. Thus the cumulative-compound motor offers more torque than a shunt-wound motor, and better speed regulation than a series-wound motor.

For a differential-compound motor, the speed depends on the difference between the two fields. Thus, at little or no load, this motor too acts

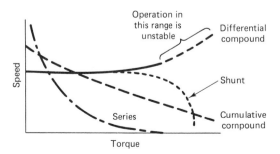

Figure 5-20   Speed/torque characteristics of compound-wound motors.

much like a shunt-wound motor. But with increasing loads, the series field comes into play and the net magnetic field is weakened. As you remember from our discussion of shunt motors, this makes the motor tend to speed up. Thus a differential-compound motor can be made to have almost perfect speed regulation; such a motor is referred to as flat-compounded.

Compound-wound motors can be reversed either at rest or during rotation by reversing the connections to the armature, as shown in Fig. 5-21.

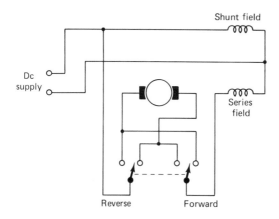

Figure 5-21 Reversing connection for compound-wound motors.

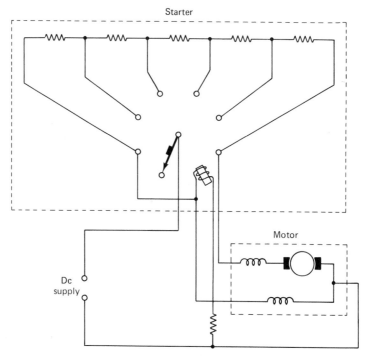

Figure 5-22 Four-point starter for compound-wound motors.

Dynamic braking can be provided by shunting the armature with a low resistance while the field coils are left connected. Control of compound motors' speed can be done either by varying the supply voltage or by a rheostat connected across the armature (as is done for series motors).

The starting current for a compound-wound motor is somewhat greater than for shunt motors, but less than for series motors. The starting current can be limited by the use of a four-point starter, as shown in Fig. 5-22.

Compound-wound motors were found in many applications in the past, in spite of their higher cost (compared to series or shunt motors). However, the availability of relatively inexpensive electronic controls has made it possible to replace them with PM or series-wound motors in many of these applications. Compound-wound motors are still used in some large dc-operated equipment in which their torque or constant-speed characteristics are desired.

SPECIAL-PURPOSE DC MOTORS

Now that we have discussed the common types of dc motors, we will look at two less common types: low-inertia and brushless designs. Low-inertia motors are designed so that their armature mass is very low. This permits them to start, stop, and change direction and speed very rapidly. These motors are useful in instrumentation applications where a very quick mechanical response to an electrical signal is required. There are two common types of low-inertia motors: *shell-type* and *printed-circuit* (sometimes called *planar coil*). In honor of its age (invented early in the twentieth century), we'll talk about the shell-type motor first. As shown in Fig. 5-23, the armature is made up of flat aluminum or copper coils bonded together to form a hollow cylinder. This cylinder is not attached physically to its iron core, which is stationary and is located inside the shell-type rotor. The absence of the iron in the rotor makes the inertia very small.

The printed-circuit (PC) motor uses a different technique to achieve low armature mass. The armature coil and the commutator segments are made in the form of a printed circuit board. The magnets of this motor are located above and below the PC armature, as shown in Fig. 5-24. The inertia of the PC armature is slightly greater than that of the shell armature. Al-

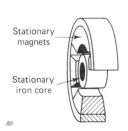

Stationary magnets

Stationary iron core

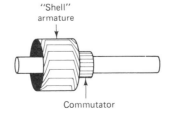

"Shell" armature

Commutator

Figure 5-23   Basic construction of a shell-type motor. (Courtesy of Bodine Electric Co.)

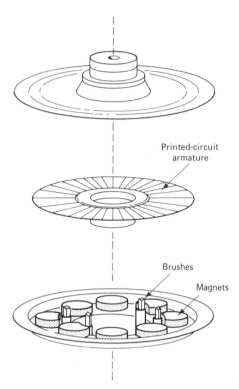

Figure 5-24   Construction of PC motor. (Courtesy of Bodine Electric Co.)

though the response is therefore somewhat slower, the torque produced is somewhat smoother, because of flywheel effect.

Both types of low-inertia motors usually use PM fields. They are rather inefficient, and are used only for very low-horsepower applications. They share the disadvantage of being able to overheat to a destructive temperature in a very short time because of the lack of iron in the armature to absorb excess heat. This makes them susceptible to burnout if they are stalled or operated with the wrong supply voltages.

Brushless dc motors are actually not dc motors at all. They are ac motors with built-in inverters to change the dc supplied to the motor into ac to be fed to the field coils. Thus the magnetic field produced by the coils rotates. The inverter's polarity-reversing circuit is triggered by sensors that sense the position of the armature in order to keep the speed constant. The armature itself is of the "squirrel-cage" design (discussed in Chapter 6). Brushless dc motors are more expensive than commutator-type designs, and are not easily adaptable to dynamic braking. These disadvantages are outweighed in some applications by the greater reliability that results from eliminating the commutator and brushes.

TABLE 5-1 Summary of DC Motor Types and Characteristics

| | | | | Motor type | | | | |
| | | | | | Compound | | | |
Characteristic	PM	Shunt	Series	Differential	Cumulative	Low inertia	Brushless
Starting torque	High	Low	High	Low	High	Low	High
Starting current	High	Low	Very high	Low	Moderately high	High	High
Reversibility	Easy	Easy	Easy	Easy	Easy	Easy	Difficult
Speed	Varying	Constant	High and varying	Very constant	Fairly constant	Varying	Constant
Speed control	Simple	Simple	Simple	Simple	Simple	Simple	Difficult
Dynamic braking	Yes	Yes	No	Yes	Yes	Yes	No
Size/weight	Smallest	Normal	Small	Large	Large	Small	Small
Cost (relative)	Low	Moderate	Low	High	High	High	High
Horsepower range	Under 1	Any	Under 2	Any	Any	Very low	Low
Uses	Portable equipment	General	Portable equipment, auto starters	General	General	Instruments	Special

MAINTENANCE

Maintenance of dc motors is discussed in Chapter 8.

SUMMARY

1. A dc motor is made up of one or more rotating (armature) coils and some means of providing a magnetic field. The more armature coils, the smoother the output torque. The polarity of the dc is reversed twice per revolution by the commutator. There are two commutator segments per armature coil. Connection to the commutator is made by graphite or copper blocks called brushes.

2. The armature coils of a motor experience an induced countervoltage that is opposite in polarity to the applied voltage. The faster the armature turns, the greater the countervoltage. The armature current is given by

$$I_{armature} = \frac{V_{applied} - V_{counter}}{R_{armature}}$$

3. The armature current results in a magnetic field that warps or skews the field produced by the motor's field magnets. Thus the neutral plane is shifted. This shifting is called armature reaction. It is compensated for by interpoles.

4. Dc motors' fields can be supplied in a number of ways, each producing a motor with its own particular characteristics. These are summarized in Table 5-1 on page 131.

QUESTIONS

1. Describe the basic principle of operation of electric motors.
2. Why is a commutator needed in a dc motor?
3. Why are electric motors made with more than two poles?
4. What does countervoltage do to the amount of current drawn by a motor? How is it related to speed?
5. What is armature reaction? How is it reduced?
6. Choose a motor for each application:
   (a) Smallest possible size
   (b) Constant speed
   (c) Maximum efficiency
   (d) Ease of speed control

7. How can a PM motor be electrically braked?

8. How does a governor in a tape-recorder motor work? What is it good for?

9. What is one major advantage and one major disadvantage of shunt-wound motors?

10. How can a shunt-wound motor be reversed?

11. What circuit can be used to control the speed of a shunt-wound motor? How does it work?

12. What happens if the field current of an unloaded shunt-wound motor is interrupted?

13. What purpose do motor starters serve? Describe the operation of three types of starters.

14. Why are series-wound dc motors so popular? Give three or more reasons.

15. What is another name for the series-wound motor? Why is this name appropriate?

16. What is the main purpose of the gear trains that are often used with series-wound motors?

17. What is the advantage of a compound-wound motor?

18. What is the difference between a differential-compound and a cumulative-compound motor?

19. What is a flat-compounded motor's special advantage?

20. What type of application are shell-type and printed-circuit motors used for?

6

AC MOTORS

AC INDUCTION MOTOR ACTION

By far, most of the motors in common use, especially in industry, run from ac power. Although many of these are of the ac-dc or "universal" series-wound variety, there is another type that gives far more reliable service. You remember that the attractiveness of a series motor's low cost is offset by the inconvenience and expense of frequent maintenance, especially brush and bearing replacement. Even if shunt- or compound-wound motors were designed for ac, with slip rings used instead of a commutator, occasional brush replacement would still be necessary. But by using the induction principle, brushes can be eliminated completely. (Yes, Virginia, there is a Santa Claus!) This principle is easier to understand if we discuss polyphase (i.e., two-or-more-phase) motors first.

You remember that a three-phase power system has ac voltages on three "phase" conductors, and those voltages are phased at $0°$, $120°$, and $240°$, respectively. This means that if a motor's field coils are wound as shown in Fig. 6-1a, the maximum magnetic field strength will behave as follows:

1. At $90°$ phase, pole A will be a strong north pole, and poles B and C will each be somewhat weaker south poles.
2. At $210°$ phase, pole B will be a strong north pole, and poles A and C will each be somewhat weaker south poles.
3. At $330°$ phase, pole C will be a strong north pole, and poles A and B will each be somewhat weaker south poles.

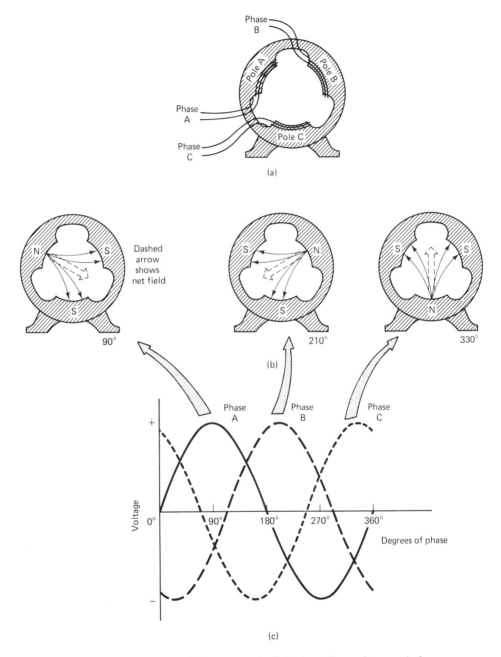

Figure 6-1 Rotation of the magnetic field in a three-phase ac induction motor; (a) connection of field windings; (b) magnetic fields at 90°, 210°, and 330°; (c) phase diagram for applied ac voltages.

Notice that the field's strong north pole is rotating; in fact, it makes one complete revolution per cycle of the applied ac. This *rotating magnetic field* is the first of two principles that make an induction motor work. The second is the principle of induced voltage. The rotor coils of an induction motor consist of only one turn each and are made up of thick aluminum or copper rods or bars, joined together at each end by an aluminum or copper shorting ring (see Fig. 6-2a). The basic structure thus resembles an exercising cage of the type used for pet gerbils. In spite of this fact, rotors of this design are called squirrel-cage rotors. Since neither a gerbil nor squirrel is really expected to live inside a motor, the space inside the "cage" is filled in with laminated iron to provide a low-reluctance magnetic path (Fig. 6-2b). Notice that in an actual rotor, the bars are angled with respect to the shaft. This produces a smoother output torque and more uniform starting performance.

When such a rotor is placed inside a field structure having a rotating magnetic field, the lines of flux of the field are moving with respect to the rotor coils (or "bars"). Thus a voltage is induced in the rotor. Because of the rotor's very low resistance, this voltage produces a high current. The high current flowing in the rotor sets up a magnetic field of its own. The magnetic interaction of the rotor field and the rotating stator field exerts a torque on the rotor, making it tend to follow the rotating magnetic field. Thus an induction motor produces a torque on the rotor without any electrical connections to the rotor. This makes the induction motor quite reliable, and hence, quite popular.

To make the induction motor principle operate with single-phase ac, the motor must incorporate some design feature that tricks the field into thinking it is two-phase. In most motors, this is done by using two separate field coils: a *start winding* and a *run winding*. The start winding is wound of smaller-gauge wire than the run winding, giving it a higher resistance. The run winding is set more deeply into the laminated iron field structure, giving it more inductance. The resulting phase difference between the two windings produces the effect of a two-phase field structure, as shown in Fig. 6-3, and

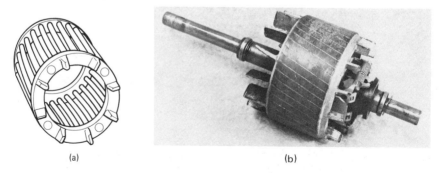

(a)   (b)

Figure 6-2   Squirrel-cage rotor construction: (a) rotor with laminations removed (courtesy of Bodine Electric Co.); (b) complete rotor.

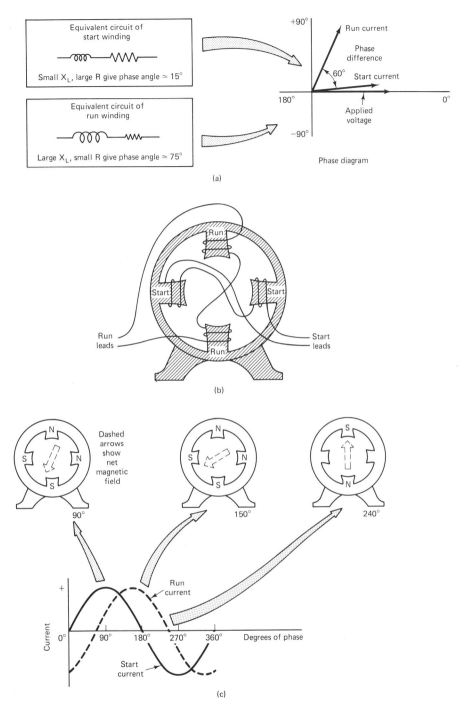

Figure 6-3 Operation of single-phase induction motor with start and run windings: (a) phase relations; (b) physical construction; (c) resulting rotating magnetic field.

the magnetic field rotates. The differences among the various types of single-phase induction motors are concerned primarily with the way in which the starting winding is connected in the circuit.

SPEED

The speed at which the magnetic field of an induction motor rotates is known as the motor's *synchronous speed*. For the motors we have discussed, the field made one complete rotation per cycle of the applied ac voltage. Assuming 60-Hz ac, we have

$$\frac{1 \text{ revolution}}{\text{cycle}} \times \frac{60 \text{ cycles}}{\text{second}} \times \frac{60 \text{ seconds}}{\text{minute}} = \frac{3600 \text{ revolutions}}{\text{minute}}$$

If the motors had more than one field coil per phase, the magnetic field would rotate more slowly. The full equation for synchronous speed is

$$\text{synchronous speed} = \frac{60f}{P}$$

where f = ac frequency

P = number of field coils per phase

Sometimes P is given in terms of "pairs of poles per phase," but since one field coil corresponds to two magnetic poles, the result is the same.

The rotor of a standard induction motor does not turn at the synchronous speed; it is slightly slower. A moment's thought will show why this is so. If the rotor bars were rotating at the same speed as the magnetic field, none of the field's flux lines would cut across the bars; that is, there would be no relative motion between the field and the rotor bars. This would mean that no voltage could be induced in the rotor, and there would be no rotor current to set up a magnetic field to supply torque to turn the rotor. The number of rpm by which the rotor speed lags behind the synchronous speed is known as the *slip* of the motor. Slip is usually expressed as a percentage of synchronous speed:

$$\% \text{ slip} = \frac{\text{synchronous speed} - \text{rotor speed}}{\text{synchronous speed}} \times 100\%$$

A typical two-pole, single-phase induction motor operated from a 60-Hz power source has a synchronous speed of 3600 rpm (as shown above) and an operating or full-load speed of 3450 rpm. The slip is then 3600 − 3450 = 150 rpm. The % slip is

$$\frac{3600 - 3450}{3600} \times 100\% = \frac{150}{3600} \times 100\% = 4.16\%$$

In case you're interested, the frequency of the ac induced in the rotor can also be calculated quite simply: rotor frequency = slip/60. The division by 60 converts from revolutions per minute to cycles per second. Thus the rotor frequency of the motor we just discussed is $150/60 = 2.5$ Hz. Rotor frequency is mentioned only as a point of interest; it has no practical importance to the motor user or service technician.

The slip of a motor is controlled by the designer; it involves such factors as the amount of air gap between the field and the rotor, the shape and resistance of the rotor bars, and the depth that the rotor bars are embedded in the iron laminations. One way in which induction motors are classified is as either *high-slip* or *low-slip*. Since most applications require a fairly constant-speed motor, the low-slip design is the most common.

Figure 6-4 shows speed/torque curves for both kinds of motors, together with types of rotors that could be used to produce motors of these types. Notice that for a low-slip motor, even though the load torque is increased, the shaft speed stays almost constant up to the overload point. When the load becomes this difficult to turn, the torque can increase no further, and the motor will stall. For a high-slip motor, the speed will vary much more as the load is changed, but it is still susceptible to stalling.

Now that we have discussed the family characteristics of induction motors, we will look at some specific types. All the motors that we will examine are of the low-slip variety, unless specifically stated otherwise.

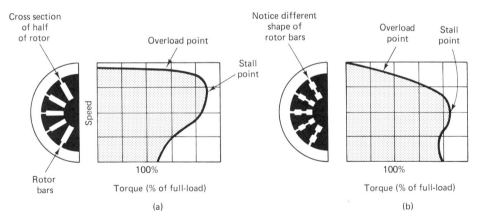

Figure 6-4 Speed/torque curves for low-slip and high-slip induction motors: (a) low-slip; (b) high-slip. (Courtesy of Bodine Electric Co.)

POLYPHASE MOTORS

The favorite motor for industrial use is the polyphase induction motor. Even though two-phase motors are occasionally found, most industrial motors are three-phase. Often, they are designed to operate on more than one supply voltage. Motors rated at 220/440 V are common, but 330- and 550-V motors are also used. Polyphase motors have the highest starting torque and lowest starting current of any standard induction motor type. They can be reversed by simply reversing the sequence of the phase connections, as shown in Fig. 6-5. Polyphase motors, or for that matter, induction motors in general, lend themselves to a simple means of dynamic braking. If dc is applied to the field coils, the resulting stationary magnetic field will induce a current in the rotor that is proportional to rotor speed. The rotor will try to "follow" the stationary field and so will quickly stop moving. As long as dc is applied to the field, the rotor will resist efforts to turn it. Circuits for braking a three-phase motor are shown in Fig. 6-6.

One disadvantage of induction motors in general is the difficulty of varying their speeds. As we mentioned earlier, the speed of an induction motor depends on its physical design and on the ac power frequency. With a normal (i.e., low-slip) motor, changing the line voltage will not change the speed significantly, although it may reduce the torque so that the motor will stall. There are, however, two methods of controlling the speed of a polyphase induction motor. The first is to change the power frequency. This requires complex and expensive electronic circuitry, especially for high-power motors. (You cannot just call up the local power company and order 45 minutes of 83-Hz power!) Also, the magnetic design of a motor is usually chosen to be optimum for a fairly narrow range of frequencies. For these reasons,

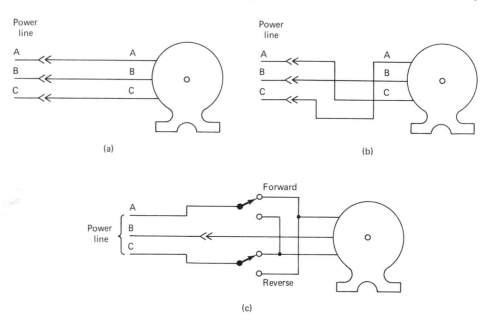

Figure 6-5 Reversing polyphase motors: (a) forward; (b) reversed; (c) switch for reversing motor.

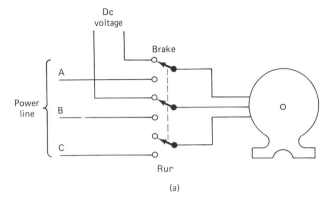

(a)

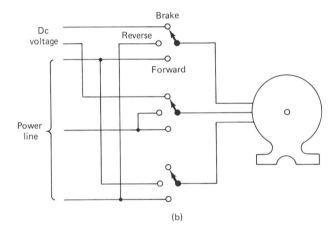

(b)

Figure 6-6 Braking circuits for
polyphase motors: (a) braking only;
(b) combined braking and reversing.

the frequency-change method of speed control is rarely used. The other
method is to change the number of poles. A three-phase motor with one coil
(two poles) per phase will turn at 3450 rpm or so, assuming 60-Hz power.
If the motor has two coils per phase, it will turn at half that speed: about
1725 rpm. Thus a motor can be made with two or three coils per phase, and
the number of coils actually connected to the power line can be switch-
selected. Thus a motor with two (3450 and 1725 rpm) or three (3450, 1725,
and 1150 rpm) speeds can be made. Unfortunately, this method makes the
motor more expensive and less efficient. For these reasons, polyphase motors that are to be used in applications requiring a variable-speed drive are
usually specially designed (i.e., high-slip or torque motors) or are connected
to a mechanical transmission that provides for the speed variations.

For the many single-speed applications in industry, polyphase induction
motors are almost the universal first choice. When a choice exists between
a single-phase and a polyphase motor, the polyphase motor will be found to
be cheaper, more efficient, and more reliable, and will have a higher starting

torque and lower starting current. Thus single-phase motors are used primarily in applications in which three-phase power is not available, as in home and light industrial applications.

SINGLE-PHASE MOTORS

Split-Phase Motors

The least expensive and most common single-phase induction motor is the split-phase motor. This motor operates on the principle discussed earlier and illustrated in Fig. 6-3. That is, the single-phase input is "split" to produce an imitation two-phase motor, with the phases about 60° apart. Because of its high resistance, the start winding would overheat if it had to pass current continuously. To avoid this, an automatic switch is used to disconnect the start winding from the ac supply as soon as the motor has reached 70% of its operating speed. The distortion of the stator's field by the rotor's field produces enough rotation of the magnetic field to keep the rotor turning. A comparison can be made with a glass of sour lemonade. You add sugar, then swirl the liquid in the glass until the sugar dissolves. If you analyze the motion of your wrist as the lemonade is being stirred, you will see that you

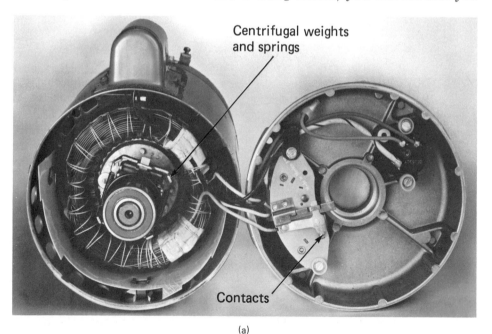

(a)

Figure 6-7 Centrifugal switches: (a) switch mounted on motor (courtesy of Century Electric, Inc.).

(b)

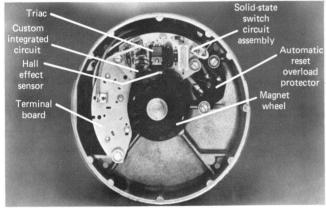

Triac
Custom integrated circuit
Hall effect sensor
Terminal board

Solid-state switch circuit assembly
Automatic reset overload protector
Magnet wheel

(c)

Figure 6-7 (b) Switch removed from motor (courtesy of General Electric Co.); (c) solid-state starting switch (courtesy of Century Electric Inc.).

make a circular motion only long enough to start the liquid swirling. You then keep it swirling with an essentially pure side-to-side motion. In the same way, once a single-phase induction motor has started turning, it can be kept turning by the "side-to-side" motion of the run windings' field, without help from the start winding. In fact, such motors can be hand-started if the start winding is defective. Once the shaft is spun, the motor will accelerate and continue to rotate in whichever direction the shaft was manually turned.

The mechanism used for automatically connecting and disconnecting the start winding is called a *centrifugal switch* or *starting switch*. It consists of two or more contact points that are held closed by spring pressure. A system of weights and levers is connected to the motor's shaft. When the shaft attains 70% of full operating speed, the weights are thrown out by centrifugal force. These, in turn, activate the levers, which push a plunger that opens the switch contacts. Photographs of two types of centrifugal switches are shown in Fig. 6-7. Oxidation of these switches is the most common cause of failure for single-phase induction motors.

In an effort to overcome the reliability problems associated with starting switches, some manufacturers have begun using solid-state devices to replace the mechanical switch contacts. The simplest of these is essentially a normally-closed thermal DOA relay that can be substituted directly for the starting switch in any motor. When power is applied, current flows for a fraction of a second to allow the motor to build up speed; then the current path is broken. These devices are available in different models for motors of various horsepower ratings and applications. They have the disadvantage that under low-line-voltage conditions, voltage may not be applied to the start winding for long enough to bring the motor up to speed. This problem is solved by a more sophisticated design that operates by sensing the power factor of the start winding. During starting, this power factor is resistive, but it becomes inductive when the motor is running normally. A coil of very few turns is connected in series with the start winding, and a reed switch is placed in the center of the coil, making a reed relay. The reed contacts are used to control the gating of a triac that either feeds or blocks current to the start winding.

The most sophisticated solid-state starting switch uses a magnet mounted on the motor shaft and a Hall-effect sensor to determine the true shaft speed of the motor. (A Hall-effect switch is essentially a solid-state version of a reed switch; it closes in the presence of a magnetic field.) When the motor speed is below 80% of rated speed, an electronic circuit causes a triac to be activated, applying current to the starting winding. This last version not only has the benefit of completely eliminating mechanical contacts, but it is the least affected by variations in line voltage and motor loading. This type of switch is illustrated in the last photo of Fig. 6-7.

Single-phase induction motors in general are supplied for either 115-V, 230-V, or dual-voltage (115/230 V) operation.

The starting torque of a split-phase motor is moderate. This means that for hard-to-start loads, other types of motors will be used. Starting current is high—8 to 10 times normal running current.

Split-phase motors cannot be reversed while in operation. They can, however, be made to start in either direction simply by reversing the connections to the start winding. This can be done via a DPDT switch.

Dynamic braking of any single-phase induction motor can be accomplished by feeding dc to the field coils, much as is done in polyphase motors. Figure 6-8 shows two methods of braking. The first is straightforward, using an external dc supply. The second uses a rectifier and resistor to charge a capacitor while the motor is running. During braking, the capacitor discharges through the motor, supplying dc to the windings.

Three cautions are necessary concerning braking of single-phase induction motors:

1. Both inductance and resistance limit the ac flow in the field windings. Only the resistance limits the dc. Therefore, a dc voltage well

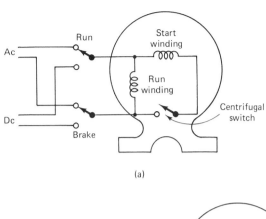

(a)

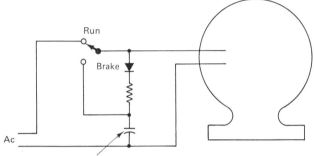

Braking capacitor of
30–200 μF for motors
up to 1/6 hp or 1000 μF
per horsepower for large
motors

(b)

Figure 6-8 Dynamic braking for
single-phase induction motors:
(a) braking with separate dc supply;
(b) capacitor-discharge braking.

below the rated ac voltage must be used to avoid overheating the
windings because of too much current.

2. Unfiltered half-wave dc supplies should not be used for braking.
   The output of such a supply may have enough 60-Hz energy to
   keep the motor turning.

3. In split-phase motors, the braking current should be applied for as
   short a time as possible, because the start winding is especially easy
   to overheat. Better still, a capacitor connected in series with the
   start winding will block dc from that winding while allowing ac to
   pass. As you will soon see, this will also somewhat improve the
   starting torque. A schematic is shown in Fig. 6-9.

The comments on speed control of polyphase motors also apply to
most single-phase induction motors. (The only exception is the PSC motor,
which is discussed later in this chapter.)

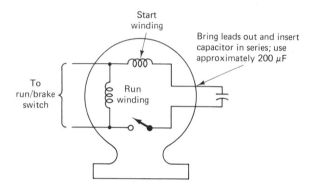

Figure 6-9 Using a series capacitor to protect the start winding during braking.

The high starting current and high start-winding resistance combine to produce quite a bit of heat during startup of a split-phase motor. In normal operation, this heat is easily dissipated since startup only lasts for a second or so. However, if a split-phase motor is heavily loaded so that the speed is kept too low for the centrifugal switch to open, the start winding will remain in the circuit, and sufficient heat may build up to damage the motor. Such overheating can also result from frequent starting and stopping of the motor.

The moderate starting torque of split-phase motors makes them unsuitable for use in compressors and other hard-to-start loads. The low cost of a split-phase motor, however, makes it a popular choice for easy-to-start applications such as fans, blowers, washing machines, and some electrically powered tools.

Capacitor-Start Motors

The reason that a split-phase motor does not have a high starting torque is the 60° phase difference between the start and run fields. A larger phase difference would produce more starting torque while slightly decreasing starting current. There is a very simple way of providing this additional phase difference. If a capacitor is connected in series with the start winding, the phase of the current in that winding will be shifted negatively. Fig. 6-10 illustrates this principle.

Except for the addition of the capacitor, a capacitor-start motor operates exactly like a split-phase motor. A centrifugal switch connects and disconnects the capacitor and start winding. Everything we said about the split-phase motor also applies to the capacitor-start motor, except for the comments on starting torque.

The capacitor itself is used for only a short period of time (usually a couple of seconds, at most) each time the motor is started. Therefore, fairly low-cost ac electrolytic capacitors are normally used. The value (μF) of the capacitor is chosen by the motor manufacturer for the best performance with that particular motor. Usually, this value is stated on the motor's name-

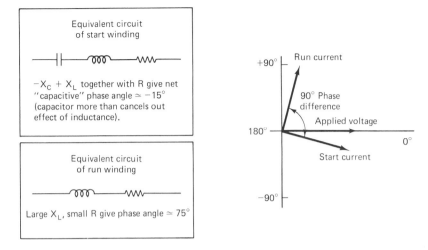

Figure 6-10 Phase relations in capacitor-start motors.

plate. The capacitor may be mounted in a "piggyback" enclosure on the motor (Fig. 6-11), or it may be mounted separately.

Capacitor-start motors are often used to turn very heavy loads, such as air compressors and refrigeration units. Therefore, they are available in large sizes. The starting current drawn by a large motor can be quite heavy. For example, a 2-hp 115-V capacitor-start motor can draw 40 A or more during startup. This much current is hard on centrifugal-switch contacts. Therefore, many capacitor-start motors rated at 1 hp or more use a *current relay* rather

(a)

(b)

Figure 6-11 Capacitor-start motors. (Courtesy of General Electric Co.)

than a centrifugal switch. The current relay's coil is made up of a few (usually 10 or fewer) turns of very heavy gauge wire. This coil is connected in series with the motor. The operate current of the relay is chosen so that the relay operates (closes the contacts) when the power is applied to the motor. As the motor's speed increases, countervoltage is generated and the current drawn by the motor decreases. At 70% of rated speed, the motor's current drops below the relay's release current level, and the relay opens. Thus a current relay does the same job as a centrifugal switch, but it opens or closes in response to the motor's current draw rather than its shaft rpm. Since the current depends on the countervoltage, which in turn depends on shaft speed, the operation of the two devices is equivalent. Current relays are often mounted in a dust-tight container located at some distance from the motor. This has the added advantage of keeping the switch contacts clean for motors that are used in dusty environments. Figure 6-12a and b show a current relay and a diagram of its application.

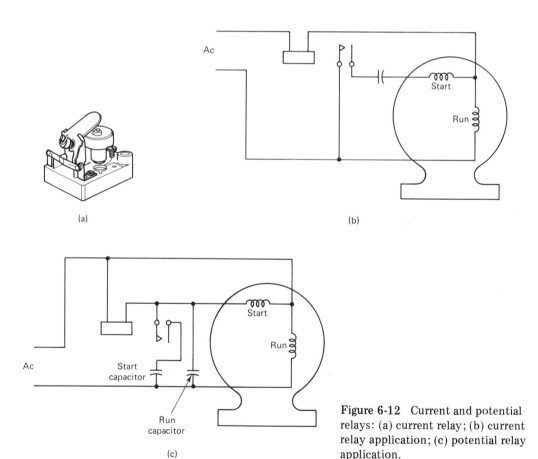

Figure 6-12 Current and potential relays: (a) current relay; (b) current relay application; (c) potential relay application.

Another device that is sometimes used for starting two-capacitor motors rather than a centrifugal switch is a *potential relay*. This device makes use of the countervoltage generated in the starting winding to operate or release the starting switch contacts. Figure 6-12c is a diagram of a motor using a potential relay. When power is first applied to the motor, the rotor is not turning, and no countervoltage is developed. Thus the full line voltage appears across the potential relay. As the motor gains speed, the countervoltage increases, so that more of the line voltage is dropped across the capacitors; the voltage across the starting winding decreases correspondingly. At 70% of the rated speed, the voltage across the starting winding becomes so small that the relay releases, leaving only the run capacitor connected.

Permanent-Split-Capacitor Motors

Since the centrifugal switch or the current relay is the most likely component of a capacitor-start motor to fail, it would be preferable if they could be left out. By using a slightly larger wire in the start winding so that it can remain connected without overheating the motor, this can be done. Of course, the capacitor must provide more negative phase shift to cancel the greater positive phase shift provided by the coil. In a motor of this type, the two windings are called the *capacitor winding* and the *main winding*. The capacitor winding is always connected, so the names "start winding" and "run winding" do not fit. Since the capacitor winding is permanently connected, and the motor is of the split-phase family, these motors are called permanent-split-capacitor motors, or PSC motors, for short. These motors operate almost exactly like capacitor-start motors, except for the following differences:

1. The starting torque and starting current are somewhat less. This is because the capacitor value is a compromise between the optimum value for starting and the optimum value for running the motor.

2. The output torque is smoother. In other words, there is less mechanical "ripple" or surging of the torque. This means lower noise and less vibration.

3. The elimination of the starting switch or relay provides more reliability.

4. The freedom from tendency to overheat the capacitor winding makes frequent starts and stops safe for the motor.

Since the capacitor used in a PSC motor is in continuous use, the more expensive oil-bath type of paper-oil capacitor must be used, rather than an ac electrolytic. The capacitor is usually not mounted on the motor itself, because these capacitors are larger for the same microfarad value than an electrolytic would be.

PSC motors cān be speed-controlled in a way that is not possible for split-phase or capacitor-start motors. By switching the series capacitor from the capacitor winding to the main winding, the torque is decreased and the slip increased. This provides two-speed operation. For this method to be effective, however, the load must be heavy enough to cause the desired increase in slip, yet not so heavy that it causes stalling.

PSC motors are often found in applications where frequent starts and stops are required, coupled with smooth, quiet operation. Examples would be reel drives ("spooling" motors) used in tape recorders and other instrumentation, and low-noise equipment fans.

Two-Capacitor Motors

To provide the smooth, quiet operation of a PSC motor and the starting torque of a capacitor-start motor, the two-capacitor-start, one-capacitor-run (*gesundheit!*) motor was invented. Although this motor's full name does describe its operation, convenience dictates the shorter name: *two-capacitor motor*. As shown in Fig. 6-13, this motor has one capacitor (the *run capacitor*) permanently connected in series with the capacitor winding. Another capacitor, called the *start capacitor*, is paralleled with the run capacitor during startup, via a centrifugal switch or current relay. This scheme provides the optimum capacitor size for both starting and running. However, it does so at the expense of bringing back the somewhat failure-prone starting switch. As you would expect, the start capacitor is usually an electrolytic, and the run capacitor is an oil-bath type. In addition to reduced noise and vibration, this motor has another advantage over capacitor-start motors. Even if the starting switch should become defective, a two-capacitor motor

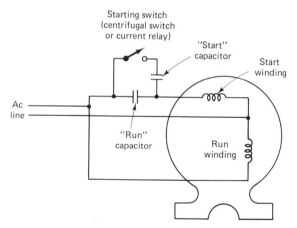

Figure 6-13   Two-capacitor motor.

will have some starting torque. This is because of the phase difference provided by the run capacitor, and the resulting rotation of the motor's magnetic field.

Shaded-Pole Motors

The operation of a shaded-pole motor is quite different from the other types of squirrel-cage induction motors. In this motor, the magnetic field is made to rotate by the inductive effect of two or more one-turn coils, or *shading* coils. Each pole of the motor's field is segmented, and the shading coil is wound around one segment. The induced current in the shading coil causes the magnetic field of the shaded pole to lag in phase behind that of the main pole. Thus the field rotates. This action is explained in more detail as follows.

1. The ac in the field coil sets up an alternating magnetic field in the poles.
2. This field induces an alternating voltage in the shading coils.
3. The induced voltage produces a large current because of the very low resistance of the shading coils. The phase of this current lags the applied voltage by 50° or so, because of the inductive effect of the shading coils. The magnetic field of the main poles thus leads the field of the shaded poles in a weak rotating action.
4. The squirrel-cage rotor follows the rotating field.

Refer to Fig. 6-14.

Shaded-pole motors have very low starting torque and are inefficient. They are not reversible unless two sets of shading coils are used, one on each side of each pole tip. Then switching must be provided to short one or the other pair of shading coils. The torque output is not smooth; it is much the same as that of a nonmotorized merry-go-round with a child turning it by successive tugs on the side bars.

With all these disadvantages, you ask, why is the shaded-pole motor ever used? The answer comes down to one word—*cheap*. The field coils of shaded-pole motors can be wound on bobbins, and that is the cheapest method of winding coils. Much less hand labor is required than for other types of motors. Also, the reliability and good speed regulation of all induction motors are also found in the shaded-pole motor. In addition, the starting current is the lowest of any induction motor. It is so low, in fact, that a stalled shaded-pole motor is not likely to overheat destructively. These characteristics make shaded-pole motors popular for small, inexpensive fans and for the cheaper models of turntables and record changers.

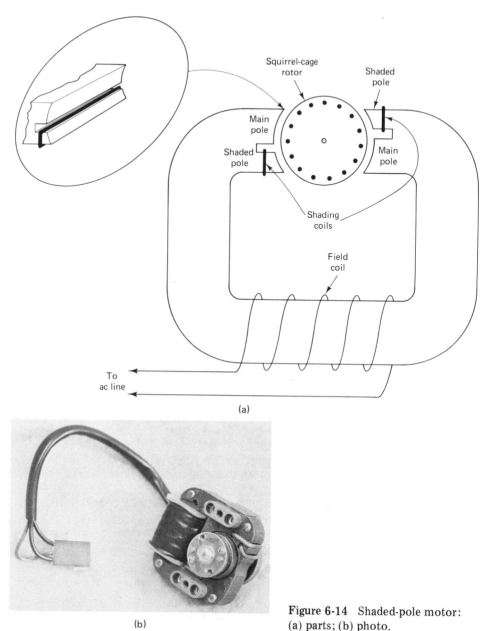

(a)

Figure 6-14   Shaded-pole motor:
(b)   (a) parts; (b) photo.

Repulsion Types

The repulsion motor is a curious combination of induction and commutator motors. The rotor is wound, just as for a dc motor. However, the power is applied only to the field coils. The brushes that connect to the commutator are shorted together. The alternating magnetic field set up by the motor's stator coils induces a voltage in the windings of the armature. This voltage produces a current, but only in those coils whose commutator segments are shorted together by the brushes. The result is a magnetic field produced by the current in those armature coils. That field is repelled by the stator field, producing a torque. The speed/torque characteristics of a repulsion motor are very similar to those of a series-wound motor: high torque at low speeds, poor speed regulation under varying loads, and a tendency to run away if lightly loaded. Since series motors require less copper, they are cheaper to build, so the repulsion motor is not commonly used.

A variation on the repulsion motor is the *repulsion-start motor*. This type of motor operates as a repulsion motor during startup. However, on reaching 70% of rated speed, a centrifugal mechanism presses a shorting ring across the commutator segments, so that the motor runs as an induction motor. For this reason, it is sometimes referred to as a *repulsion-induction* motor. No matter what you call them, such motors are seldom used today, since their only claims to fame—high starting torque and good speed regulation—are shared, at lower cost, by capacitor-start motors. Also, the absence of brushes in capacitor-start motors makes their reliability greater than that of a repulsion-start motor.

SPEED CONTROL OF AC MOTORS

The primary methods of speed control that are applicable to ac motors have already been discussed. However, the most popular method of speed control—electronic control—has not been described yet. Electronic speed control is limited to use with series-wound motors. However, since variable-speed motors of this type comprise the vast majority of variable-speed ac motors, we will take some time to look at how these controls operate.

You remember from Chapter 3 that there are a couple of electronic devices that act very much like latching relays: namely, the SCR and the triac. These devices can be used to turn the ac power to a motor on for only a part of each cycle, reducing the *average* voltage to the motor. Since the devices act as switches, they do not have to dissipate large amounts of power as a rheostat would if it were used for speed control.

The basic principle on which most speed-control circuits work is the use of a phase-shifted gating signal to control the SCR or triac. "Phase-shifted" means that the gating signal is not in phase with the voltage applied to the motor. The operation of a simple triac speed control is explained in Fig. 6-15. Since this type of speed control depends on the potentiometer-controlled phase shift of the gating signal, it is called a *phase-control* circuit. Just remember that the name is not meant to imply that the circuit controls phase; it probably should be called a phase-controlled circuit. This circuit

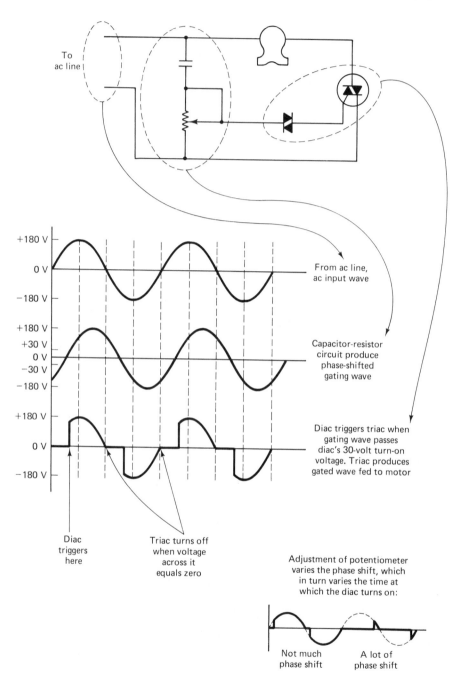

+180 V

0 V

−180 V

From ac line,
ac input wave

+180 V
+30 V
0 V
−30 V

−180 V

Capacitor-resistor
circuit produce
phase-shifted
gating wave

+180 V

0 V

−180 V

Diac triggers triac when
gating wave passes
diac's 30-volt turn-on
voltage. Triac produces
gated wave fed to motor

Diac
triggers
here

Triac turns off
when voltage
across it
equals zero

Adjustment of potentiometer
varies the phase shift, which
in turn varies the time at
which the diac turns on:

Not much
phase shift

A lot of
phase shift

Figure 6-15   Triac motor speed control.

can also be used with an SCR rather than a triac. The output wave would then be half-wave rectified; it would consist of positive or negative half-cycles only, since the SCR passes current in only one direction.

The circuit shown in Fig. 6-15 has the disadvantage of reducing the torque as well as the speed. There is a refinement of this circuit that compensates for this effect. Figure 6-16 has the motor connected between the triac's anode 1 and gate. Thus when the motor's countervoltage is large (high speeds), the gate-to-anode 1 voltage is reduced, and the ac is turned on for less of each half-cycle. When the contervoltage is small (low speeds), the triac passes more ac energy to the motor, giving it more torque to pull the speed up.

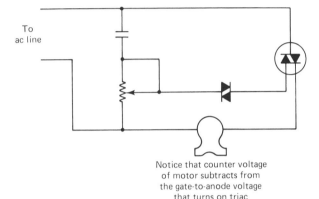

To
ac line

Notice that counter voltage
of motor subtracts from
the gate-to-anode voltage
that turns on triac

Figure 6-16 Triac speed control
with better speed regulation.

An unfortunate side effect of SCR or triac motor speed controls is their tendency to generate large amounts of radio-frequency (RF) noise. This noise can be reduced by a low-dc-resistance RF choke inserted in series with the ac line, and a capacitor (0.1 μF or so) connected across the line. For serious RF interference cases, power-line RFI filters are available.

VARIATIONS

High-Slip Motors and Torque Motors

Often, an ac motor is required that has the reliability of an induction motor, but the speed-control ability and freedom from stalling of a series motor. By increasing the air gap and/or changing the design of the stator and rotor, an induction motor can be designed to have unusually high slip. Such motors are somewhat less efficient than the normal low-slip designs, but otherwise the performance is similar. Figure 6-17 shows the speed/torque characteristics for a standard low-slip motor, compared to those for a high-slip motor.

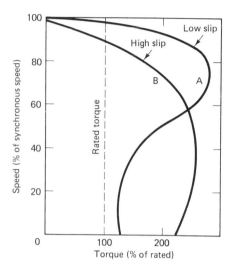

Figure 6-17   Comparison of speed/torque curves for low-slip and high-slip induction motors. (Courtesy of Bodine Electric Co.)

A high-slip motor such as this can be slowed to about 40% of rated speed before the stall point is reached. This slowing can be achieved by reducing the supply voltage, increasing the load, or a combination of the two. Thus a triac speed control could be used successfully with this type of motor, even though it would not work with a normal induction motor.

There are other applications that call for an extremely high-slip motor—one with a speed/torque curve much like that of a PM motor. You remember from Chapter 7 that a PM motor's torque increases linearly as speed decreases. Of course, a series-wound, compound-wound, or PM-type motor could be used. However, all three of these have commutators and brushes, making them noisier and less reliable than induction motors. By carrying the high-slip design to extremes, an induction motor can be designed that has the desired characteristics. Such motors are called *torque motors*. Figure 6-18 shows the speed/torque characteristics of a torque motor compared to those of low-slip and high-slip inductor motors. Notice that the torque is nearly maximum at zero speed. In fact, the rated torque of a torque motor is its *stalled-rotor torque.*

The uses for torque motors fall into three classifications:

1. *Tensioning:* These are applications in which the motor acts more or less like a spring to maintain a certain amount of tension; no rotation is required. Unlike a spring, the amount and direction of force supplied by a torque motor can be controlled via its supply voltage and connection.

2. *Actuators:* These are applications in which a rotary motion of a few degrees or a few revolutions is needed to open or close a device such as a valve.

3. *Spooling:* These applications are best represented by the reel drive motors of a reel-to-reel tape recorder. During play, the take-up reel motor must turn very slowly, its speed varying as the amount of tape on the reel

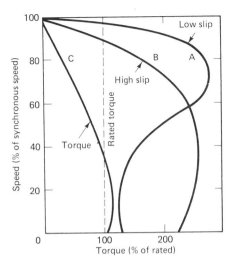

Figure 6-18 Speed/torque curves for torque motors compared with those for low-slip and high-slip designs. (Courtesy of Bodine Electric Co.)

changes. The supply reel motor must turn against its normal direction of motion, and its speed, too, must vary. Both motors' torques must remain essentially constant. During fast-forward or rewind, the take-up reel motor must turn at a high speed, while the supply reel motor must again supply reverse torque.

The most common types of ac torque motor are the two- or three-phase motor and the PSC single-phase type. The speed and torque are controlled by varying the supply voltage. Often, when a PSC torque motor is used, full voltage is applied to the capacitor winding, while the voltage applied to the main winding is varied. Thus a stable torque can be maintained even at low speeds. Figure 6-19 shows the effect of varying supply voltage upon a motor's speed/torque curve.

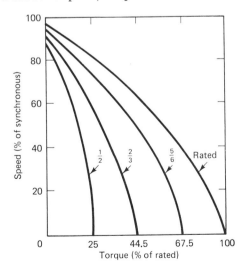

Figure 6-19 Variation of torque motor's speed/torque curve with changes in supply voltage. (Courtesy of Bodine Electric Co.)

Because of the low (or zero) operating speed of torque motors in many applications, external ventilating fans driven by other motors are often required to keep the torque motor cool.

Synchronous Motors

All the induction motors that we have discussed so far—low-slip, high-slip, and torque motors—have had *some* slip. The last variety of ac motor we will discuss has no slip; it turns at the synchronous speed. Such *synchronous motors* are used in timing applications, turntables (at least the older ones), and tape drive systems, where a motor of constant and predictable speed is essential. There are two common types of synchronous motors. Each is named for the principle of its operation.

The *reluctance-synchronous* motor uses a rotor that is slightly different from that of a normal induction motor. This difference consists of notches in the laminated portion of the rotor. The resulting four-leaf-clover-shaped rotor acts very much like two crossed bar magnets. The poles of these "magnets" (called *salient poles*) lock in with the motor's rotating field, turning at exactly the same (synchronous) speed. The name "reluctance-synchronous" comes from the additional reluctance introduced into the magnetic circuit by the notches between the salient poles. Figure 6-20 shows a diagram and a photograph of the rotor of a reluctance-synchronous motor.

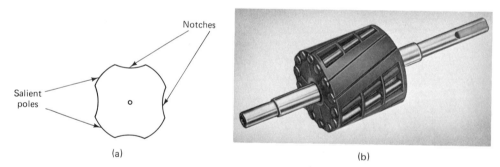

(a)   (b)

Figure 6-20 Construction of the rotor of a reluctance-synchronous motor: (a) rotor laminations (cross section); (b) complete rotor. (Courtesy of Bodine Electric Co.)

During startup, a reluctance-synchronous motor behaves exactly like any other induction motor. They may be of the multiphase, split-phase, capacitor-start, or PSC types. As the motor reaches approximately 95% of synchronous speed, however, a very different thing occurs: The rotor will suddenly snap into synchronism. This is illustrated in Fig. 6-21. At the "pull-in" point, the motor's torque is somewhat lower than it is just below that point. There is therefore a critical amount of torque that the reluctance-

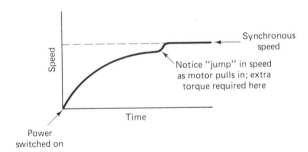

Speed

Synchronous
speed

Notice "jump" in speed
as motor pulls in; extra
torque required here

Time

Power
switched on

Figure 6-21   Pull-in characteristic
of reluctance-synchronous motors.

synchronous motor can supply and still pull into synchronous speed. This is called the motor's *pull-in torque*. There is also a maximum torque that the motor can supply while at synchronous speed without being pulled out of synchronism. This is called the *pull-out torque*. When the motor is required to turn a load requiring more torque than the pull-out torque, it will operate as a normal induction motor, except that the speed will be very irregular, resulting in a pounding noise as the motor speeds up and slows down. Provided that it is not thus overloaded, the reluctance-synchronous motor operates at an extremely stable, accurate speed.

The other type of motor uses a rotor that doesn't even look as if it should work! It is made of a hollow cylinder of a hard ferromagnetic material (see Fig. 6-22). The motor's magnetic field induces magnetic poles in the rotor. These lock in with the rotating magnetic field in much the same way as do the salient poles in a reluctance-synchronous motor. The poles, once induced, tend to stay in more or less the same position on the rotor, because of hysteresis of the rotor material. Of course, this is impossible until the rotor reaches synchronous speed. Once at synchronous speed, however, it is

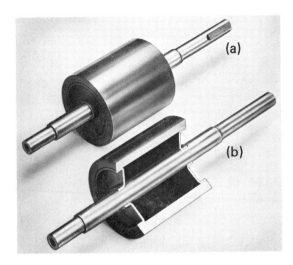

(a)

(b)

Figure 6-22   Construction of the
rotor of a hysteresis-synchronous
motor. (Courtesy of Bodine
Electric Co.)

this tendency that keeps the rotor in synchronism: thus the name *hysteresis-synchronous* motor.

Hysteresis-synchronous motors can be started by any of the methods that apply to other induction motors. They start and pull up to speed more gradually than do reluctance-synchronous motors, and they also pull into synchronism gradually (see Fig. 6-23). That is, they do not have a critical pull-in or pull-out torque, nor is their speed as unstable below synchronism as that of a reluctance-synchronous motor. These advantages are somewhat offset by the higher cost of these motors.

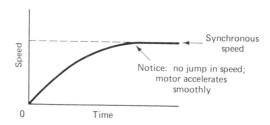

Figure 6-23 Pull-in characteristic for hysteresis-synchronous motors.

Synchronous motors of either type are more expensive than their non-synchronous counterparts. They are also somewhat larger and heavier for a given horsepower output. For these reasons, synchronous motors are usually used only when their extremely accurate speed characteristics are essential.

SUMMARY

1. Multiphase induction motors operate because of:
   a. rotating magnetic fields produced by the stator coils, that
   b. induce voltages in the rotor, causing
   c. current to flow in the rotor, setting up
   d. a magnetic field that causes the rotor to attempt to follow the rotating magnetic field produced by the stator.

2. The rotor of most induction motors is called a squirrel-cage rotor because of the appearance it would have if the laminated iron were removed.

3. Single-phase induction motors require some method of "artificially" making the stator's magnetic field rotate. This can be done by means of a starting winding that is physically located in between the main stator poles, and whose current is phase-shifted from the main winding's current by resistive, inductive, and/or capacitive effects.

TABLE 6-1 Summary of AC Motor Types and Characteristics[a]

Motor type	Starting torque	Starting current	Reversibility	Relative cost	Horsepower range	Uses
Multiphase	High	Medium	Easy, at rest or in motion	Low-normal	Any	General industrial
Split-phase	Moderate	High	Easy, at rest	Normal	Up to 2	Fans, washers
Capacitor-start	High	Medium	Easy, at rest	High-normal	Up to 5	Compressors, power tools
PSC	Moderately high	Medium low	Easy, at rest or in motion	High-normal	Up to 5	Tape recorders
Two-capacitor	High	Medium	Easy, at rest	High-normal	Up to 5	Compressors, power tools
Shaded-pole	Low	Low	Not reversible	Low	Up to $1/2$	Fans, turntables
Repulsion	High	High	Easy, at rest	High	Any	Very few nowadays

[a]Other characteristics include:

1. *Speed*: relatively constant except for torque motors (varying) and synchronous motors (extremely constant).
2. *Speed control*: two- and three-speed designs possible; otherwise, speed control difficult except for high-slip and torque motors.
3. *Dynamic braking*: yes, all types; exercise care to avoid burning out windings.

161

4. An induction motor's magnetic field rotates at the motor's synchronous speed:

$$\text{synchronous speed} = \frac{60 \times \text{line frequency}}{\text{number of field coils per phase}}$$

The rotor of most motors turns somewhat more slowly. How much more slowly is called slip, and is usually expressed as a percentage.

5. Squirrel-cage induction motors operate at a relatively constant speed unless too heavy a load is applied. With most motors, an overload will decrease speed only slightly before causing the motor to stall.

6. Most industrial motors are of the three-phase variety.

7. Split-phase motors are single-phase motors that use two windings constructed differently to produce an electrical phase difference. The start winding is disconnected by a centrifugal switch when the motor is operating at full speed.

8. Capacitor-start motors use a start winding and a capacitor to produce phase shift, giving greater starting torque and lower starting current. They also use either a centrifugal switch or a current relay.

9. Permanent-split-capacitor (PSC) motors have a capacitor permanently connected in series with one winding. No centrifugal switch or current relay is required. Their output torque is smooth and their operation is quiet.

10. Two-capacitor motors combine the PSC motor's quiet operation with the starting characteristics of a capacitor-start motor.

11. Shaded-pole motors use a segmented pole tip partially surrounded by a shading coil to produce a rotating field. They are inexpensive, inefficient, and have very low starting torque.

12. Reluctance-synchronous and hysteresis-synchronous motors are special types of induction motors that operate at the synchronous speed.

13. Table 6-1 (page 161) compares the performance of the various types of motors.

QUESTIONS

1. Why do induction motors not require commutators?
2. What is the angular spacing of the windings in a two-phase induction motor?
3. What two principles are involved in the operation of an induction motor?

4. Why does a split-phase motor require a starting switch or relay while a polyphase motor does not? What other types of motors use starting switches?

5. How does a start winding differ from a run winding? Why is this difference important?

6. True or false: An induction motor always rotates at its synchronous speed.

7. True or false: Motors having more coils and more phases will rotate more slowly.

8. What is the synchronous speed of a 400-Hz three-phase three-coil motor?

9. What is slip? For what purpose are high-slip motors made?

10. How can a three-phase motor be reversed?

11. How can a split-phase motor be started if the starting switch is defective?

12. List in order of starting current beginning with the lowest: two-capacitor motors; split-phase motors; polyphase motors; capacitor-start motors; PSC motors.

13. List the motor types given above in order of starting torque, beginning with the lowest.

14. How can a single-phase induction motor be electrically braked?

15. What three cautions apply to braking of single-phase motors?

16. What advantage over a split-phase motor are offered by:
    (a) Capacitor-start motors?
    (b) PSC motors?
    (c) Two-capacitor motors?

17. How does a current relay work, in conjunction with a capacitor-start motor?

18. What method of speed control can be used for PSC motors, but not for other kinds?

19. How does a shaded-pole motor operate? Why are so many used?

20. What type of ac motor can be speed-controlled by a triac circuit? Describe how such a circuit works.

21. Name two methods of reducing RFI generated by motor speed controls.

22. What type of motor is used when extremely high slip is required?

23. Why do many reel-to-reel tape recorders have ventilating fans for cooling the reel motors?

24. What is a zero-slip motor called? What are two types of such motors?

25. Define pull-in torque and pull-out torque.

7

OTHER ROTATING
ELECTRICAL MACHINERY

The ac and dc motors and generators we discussed in the last three chapters comprise the most common types. However, there are several important types of rotating electrical machinery that do not quite fit in with these for one reason or another. These devices either are not exactly motors and generators, or they are motors-plus-generators, motors-plus-control devices, and so on. We will look at several such animals now.

MOTOR-GENERATORS

A motor-generator is—guess what!—a motor and a generator mechanically coupled together. Its job is to convert electrical energy to mechanical energy and back to electrical energy. Before you begin making remarks about the motor-generator obviously being a government invention, I should add that the output electrical energy is in a different form from the input electrical energy. Motor-generators may be dc-in/ac-out or ac-in/dc-out or single-phase-ac-in/three-phase-ac-out or low-voltage-dc-in/high-voltage-dc-out or . . . well, you get the picture. At one time motor-generators were quite common ways of converting battery power (12 or 24 V dc) to high-voltage ac for operating transmitters, receivers, and so on. Now that solid-state electronic devices can do the same job more cheaply, efficiently, quietly, and reliably, motor-generators are becoming less and less common.

The output voltage of a motor-generator is often made adjustable by means of a rheostat in series with the generator's field coil. Some large dc-input motor-generators use starters to limit the initial current to the motor. The load is never connected to a motor-generator until it has reached full operating speed, because of the extra current that the motor would then be likely to draw.

Any motor-generator that uses brushes is liable to a certain amount of sparking, which causes radio-frequency interference (RFI). For this reason, brushes are often connected to each other and to the chassis by small capacitors. The capacitors significantly reduce the production of interfering radio waves.

DYNAMOTORS

One bright day, it struck someone that there was no need for a motor-generator to have two shafts, two sets of bearings, two field windings, and sometimes even two separate housings. So he designed a motor-generator that had two sets of armature coils, but only one shaft and field coil. This device is called a dynamotor. As you would expect, it is smaller, lighter in weight, and less expensive than a standard motor-generator. Also, as you would expect, there are drawbacks. First, dynamotors are all dc-to-dc. Second, the output voltage cannot be adjusted easily. Third, because field design cannot be optimized for both the motor and the generator armatures, a compromise is made, resulting in lower efficiency. Finally, the voltage regulation, or ability to maintain a constant output voltage with variations in load current, is only fair. Solid-state devices have almost driven the dynamotor into oblivion.

DC STEPPER MOTORS

Often, it is useful to have a device that will rotate only a specific fraction of a revolution—such as $15°$—on command. Such a device could control the motion of its load very accurately. These devices exist; they are called stepper motors. For example, a numerical-controlled (NC) drill is a special drill press whose worktable can be moved forward, backward, left, or right by means of programs punched onto paper tape. The drill head can be made to descend by a chosen number of mils (thousandths of an inch) by the same method. By use of an NC drill, thousands of workpieces can be drilled very accurately according to the same pattern, without direct human control. The positioning of the worktable and drill head of an NC drill are accomplished by stepper motors.

How does a stepper motor work? Well, do you remember that an ac induction motor can be braked by applying dc to the windings? Basically, the induced magnetic field in the rotor tries to follow the stator field. Rotation is produced by causing the stator field to rotate. Braking results when the stator field is made stationary, so that the rotor "follows" that stationary field. If a short burst of ac is applied to make the field rotate, dc is applied again, the rotor will turn, then stop again. Unfortunately, the amount of rotation of the rotor could not be very accurately determined; it would depend upon the duration of the applied current and upon the load. By special design of the stator's magnetic circuit, and by application of a very specific waveform, stepper motors can provide very accurate amounts of rotation. Here's how they do it. The stator is made up of two parts. Each part consists of a circular coil with a stamped-iron pole piece on top and bottom. The pole pieces themselves have alternately located teeth, as shown in Fig. 7-1a. Figure 7-1b shows the alternating north and south poles that result from this arrangement. The two identical stator assemblies are positioned so that the pole pairs of the bottom assembly are slightly out of alignment with those of the upper assembly. The amount of the misalignment is one-fourth of the

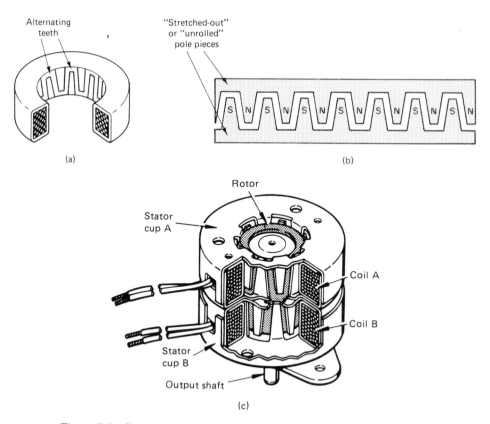

Figure 7-1 Construction of a stepper motor: (a) part of stator assembly; (b) magnetic north and south poles formed by pole-piece teeth; (c) cutaway view of assembled motor (drawing reprinted by permission, courtesy of Airpax Corp.).

distance between poles. The result of all this is that if the relative polarities of the two coils are changed, the magnetic field rotates by one-fourth of the distance between poles. For example, if each stator assembly has 12 north and 12 south poles, the distance between poles is $360°/12$, or $30°$. Reversing the direction of current through one coil (hence reversing the relative magnetic field polarities) causes the field to rotate by one-fourth that amount, or $7.5°$. Forty-eight such steps would complete one revolution.

So far, we have talked only about the rotation of the magnetic field. What about the rotor? Well, it follows the rotation of the field. When the field becomes stationary, the rotor is magnetically held stationary. The cheapest kind of rotor used in stepper motors is the *reluctance rotor*, which is a toothed, soft-iron rotor as depicted in Fig. 7-2. The teeth are responsible for focusing the magnetic field lines through the rotor, making accurate stepping action possible. A more accurate method is to use a permanent-magnet (PM) rotor. If PM rotors are toothed, even greater accuracy and higher torque will result. Stepper motors, then, are available in three types: reluctance, PM, and PM-hybrid, the latter being the one that uses the toothed PM rotor.

Figure 7-2 Rotor construction used in reluctance and PM-hybrid stepper motors.

The most important characteristics of stepper motors are the step angle, holding torque, residual torque, dynamic torque, operating voltage, and winding type. The *step angle* is simply the number of degrees of rotation per step. Typically, this is either $7.5°$ (48 steps/revolution), $15°$ (24 steps/revolution), or $18°$ (20 steps/revolution). The step angle is usually accurate within less than $\pm 10\%$. Any error in step angle is canceled out in future steps, so that a $15°$ stepper motor will position accurately to within $1.5°$ (10% of $15°$), regardless of the number of steps. Thus one step would be $15° \pm 1.50°$, and 1000 steps would be $15,000° \pm 1.5°$.

The *holding torque* is the torque with which the motor holds its position in between steps. The *residual torque* is the holding torque for a nonenergized PM or PM-hybrid stepper motor (holding torque with power cut off). The *dynamic torque* is the turning force the stepper motor can exert on a load. It is usually less than the holding torque. The dynamic torque decreases as the motor speed is increased, as shown in Fig. 7-3.

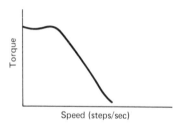

Speed (steps/sec)

Figure 7-3 Speed/torque curve for a typical stepper motor.

Stepper motors are available in virtually any desired operating voltage. The rated operating voltage will produce the rated dynamic and holding torques. Greater voltage will produce more torque, but may cause the motor to overheat because of excessive current. Too low a voltage will result in torque being below the rated value.

Stepper motors are available with two winding types: two-coil bipolar and four-coil unipolar. These winding types are diagrammed in Fig. 7-4a.

Obviously, the number of steps produced by a stepper motor is not controlled by a tiny little man with a polarity-reversal switch. Instead, electronic circuits are used to produce a certain number of pulses of a square wave. There are three basic drive methods that can be used to power a stepper motor. The four-step sequence is the most common. The eight-step sequence is sometimes used, and gives twice as many step increments (i.e., $3.75°$ steps from a $7.5°$ stepper motor). The wave drive method energizes only one coil at a time, giving greater efficiency. However, the step accuracy of either an eight-step or a wave-drive system is poorer than that of the four-step system. These three drive methods are illustrated in Fig. 7-4b. Notice that no matter which sequence is used, the relative direction of current in the coils reverses once per step. For the unipolar drive, the same sequences can be used, but only one transistor is turned on at a time.

It was mentioned earlier that the torque of a stepper motor decreases as motor speed is increased. This is mainly because of the inductance of the windings, which reduces the current that can flow in them as frequency (number of pulses per second) of the drive waveforms is increased. There are three methods of partially compensating for this effect. The *L/4R drive* circuit shown in Fig. 7-5a uses a resistor connected in series with the center-point lead of each pair of windings. This resistor is equal to three times the motor's winding resistance. Four times the rated voltage is applied, and at zero speed, the motor is fed one-fourth of this total (i.e., its rated voltage). At high speeds, the inductive reactance of the coils increases, and proportionally more of the total voltage is applied to the motor, resulting in greater current in the windings: hence greater torque.

The *bilevel drive* circuit shown in Fig. 7-5b applies voltage from the low-voltage supply to provide holding torque. When the motor is to be made to step, transistors Q_1 and Q_2 apply voltage from the high-voltage supply to give greater dynamic torque.

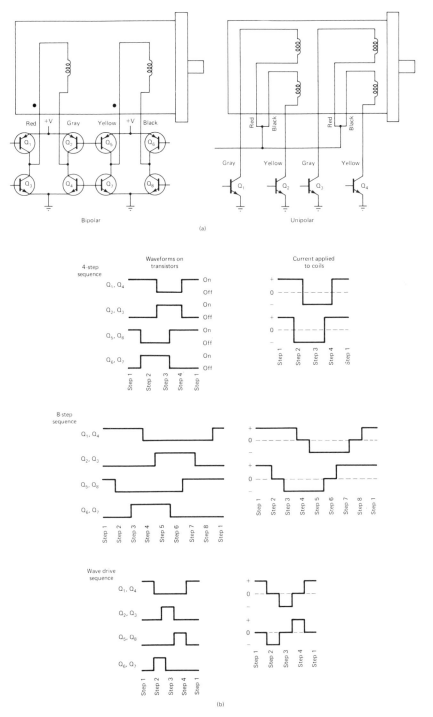

Figure 7-4 Drive methods for stepper motors: (a) winding types; (b) drive methods.

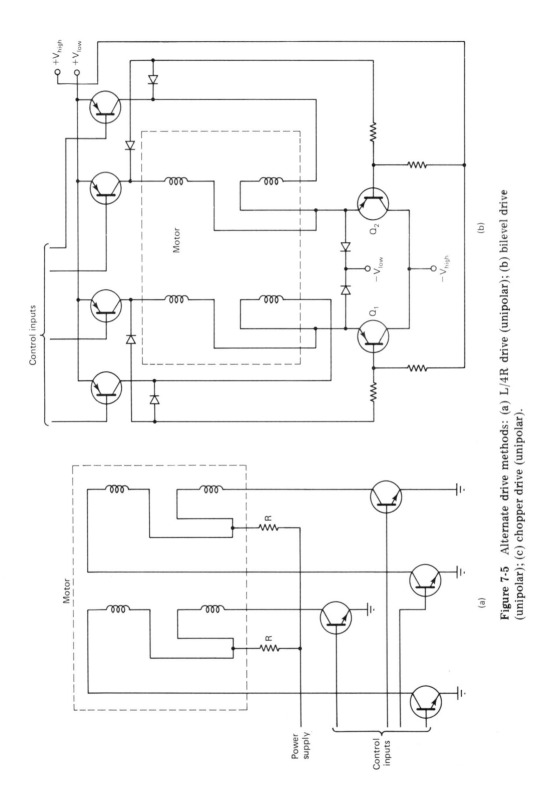

Figure 7-5 Alternate drive methods: (a) L/4R drive (unipolar); (b) bilevel drive (unipolar); (c) chopper drive (unipolar).

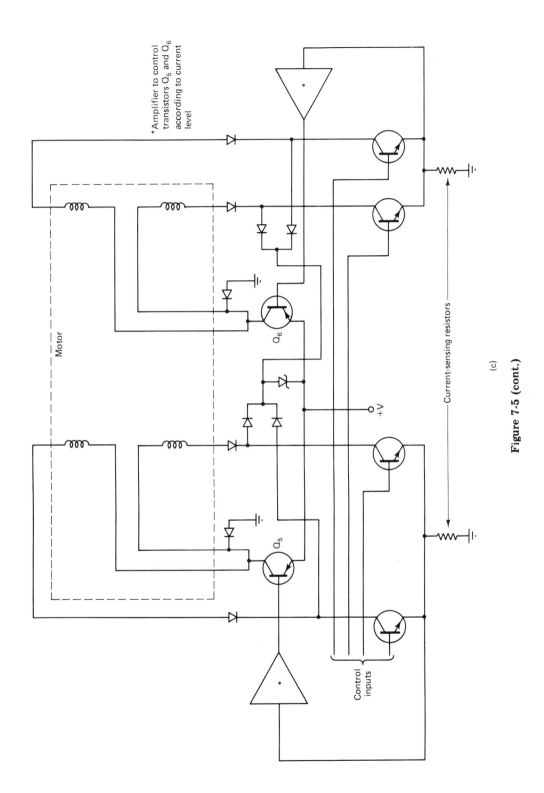

*Amplifier to control transistors Q_5 and Q_6 according to current level

Motor

Control inputs

Current-sensing resistors

+V

Q_5

Q_6

(c)

Figure 7-5 (cont.)

The *chopper drive* circuit in Fig. 7-5c uses an electronic sensor and amplifier to maintain constant coil current and dynamic torque.

All three of these circuits attempt to maintain a more constant dynamic torque by maintaining a more constant coil current at all drive frequencies. But for every motor there is a limit beyond which more current will not improve the torque significantly. Thus for high-speed applications, one more specification becomes important: the *maximum number of steps per second.*

The transistors shown in Figs. 7-4 and 7-5 are not a part of the motor; they are the final drive components of the electronic "tiny little man with a switch." Notice that no connections are shown to the base leads. These are the signal input leads from the control circuitry. The control circuitry itself determines the number and timing of pulses to be applied to the base leads of the final drive transistors. The operation of this type of circuit is covered in courses in digital electronics and is beyond the scope of this book.

When a stepper motor is operated without a load, as might be done when testing the motor, two phenomena can show up that can be disturbing to the technician who does not know to expect them. The first is *ringing.* When a nonloaded stepper motor is made to step and then halt, it will somewhat overshoot the proper stopping place, then bounce back and forth as it gradually settles into position. The ringing occurs quite rapidly, but it can produce noises rather as though something were loose in the motor. The second phenomenon is *resonance.* At certain speeds, a stepper motor will resonate mechanically, producing noticeable vibration and/or noise that can make the unwary technician suspect a faulty bearing. Both of these phenomena are greatly reduced or eliminated when the motor is operated with a load.

Sometimes a device is needed that will push or pull a control, but with more precision than a solenoid offers. By combining a stepper motor with a precisely threaded shaft, the *digital linear actuator* was invented. This device provides very accurate linear motion controllable to within 0.001 in. Digital linear actuators are available with a throw varying from 1/2 to 2 1/2 in. and with a force up to 25 lb.

GEARMOTORS

Except for stepper motors, none of the motors we have discussed is suitable for operation at speeds below several hundred rpm. And stepper motors are inefficient, require complex driving circuitry, and move in "jerks" rather than rotating smoothly. Second, there are many applications in which a shaft must be turned at 1000 to 3000 rpm—quite reasonable for an induction motor—but in which small size and weight are required, dictating the choice of a series-type or PM motor. Finally, there are a number of applications in which a large torque is required from a compact motor. All three of these types of applications can be filled by a series or PM motor having a built-in gear train. Other types of drive motors can be used, but series and PM are the most common. The function of the gear train is twofold: to reduce shaft speed and to increase output torque. Figure 7-6 shows three varieties of gearmotors.

Gearmotors can be obtained with virtually any combination of horsepower and shaft speed desired. They require somewhat more maintenance than do nongeared motors, as the gear trains require occasional lubrication. They are also noisier and about 3 to 15% less efficient.

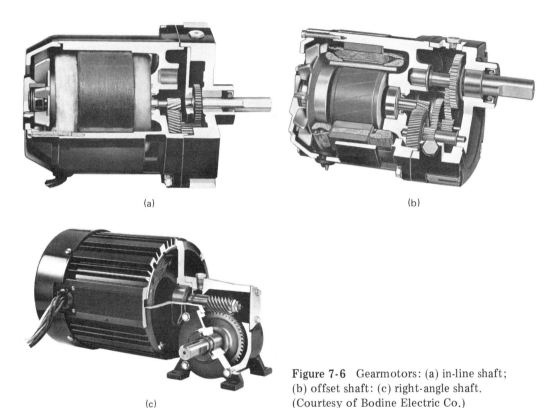

(a)

(b)

(c)

Figure 7-6   Gearmotors: (a) in-line shaft; (b) offset shaft: (c) right-angle shaft. (Courtesy of Bodine Electric Co.)

SERVOMOTORS AND SYSTEMS

Sometimes the speed regulation required from a motor is greater than any simple motor can deliver. In some such cases, the speed must be capable of being controlled very accurately. Since all the motors we have discussed have some variation of speed as the load is varied, something more is needed. That something is a servo system. A typical servo system is diagrammed in Fig. 7-7.

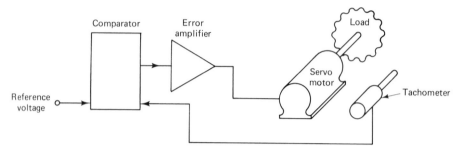

Figure 7-7 Block diagram of a servo system.

It works as follows. The servomotor itself turns the load. The speed of the load is sensed by a *tachometer*, which produces an output voltage proportional to the motor speed. The output from the tachometer is fed to one input of a *comparator*, which is a circuit whose output depends on the difference between two input voltages. The other input voltage to the comparator is a *reference voltage*—the same voltage level that the tachometer produces when the shaft speed is correct. If the speed is right, the two input voltages are equal and the comparator's output is zero. If the speed is too high, the tachometer's voltage is higher than the reference voltage, and the comparator puts out a negative voltage. If the speed is too low, the tachometer's voltage is lower than the reference voltage, and the comparator puts out a positive voltage. The output voltage from the comparator is called the *error voltage*. The error voltage is fed to an *error amplifier*, which puts out a voltage to drive the servomotor. Thus if the shaft speed is incorrect, the servo system automatically changes the drive voltage fed to the motors, so that the speed is corrected.

Integrated-circuit technology has made it possible to use a slightly different, but more accurate type of servo system in which the tachometer produces an ac output whose frequency varies with the shaft speed. This frequency can then be compared to a reference frequency to derive the error signal. Using this system, speed variations can be held to less than one-hundredth of a percent. A common example of this type of servo system is the drive motor for many high-quality home cassette decks. Usually, replacement of a motor in such a unit also necessitates replacement of the motor-speed-control board, which incorporates the frequency comparator and error amplifier (see Fig. 7-8).

The motors used with servo systems may be of many different types. The most common are PM (dc) and single-phase PSC or two-phase (AC). PM motors are well suited to servo application, since their speed is directly proportional to the applied voltage. Ac servomotors must be of the high-slip variety; otherwise, varying the input voltage will not change the speed.

Figure 7-8 Dc servomotor and control board.

The two windings of an ac servomotor are connected as shown in Fig. 7-9. The main winding (called the *fixed phase*) is usually connected to 115- or 230-V ac. The *control phase* can be designed for 15- to 36-V, 115-V, or 165-V operation. For applications requiring rapid changes in speed or direction, *low-inertia* servomotors may be used. These motors have smaller rotors than standard motors, so their mass is less, making for quicker acceleration and deceleration. Low-inertia motors are slightly less efficient than standard motors, however.

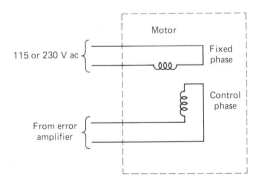

Figure 7-9 Ac servomotor connection diagram.

SYNCHROS AND RESOLVERS

In the children's game called "Follow the Leader," the object is for each participant to exactly mimic the motions of the leader. Synchros are electromechanical Follow-the-Leader players. A common application for a synchro is in a ship's radio antenna system. A synchro transmitter is attached to the antenna so that when the antenna is rotated (by a gearmotor), the synchro receiver in the control room follows the rotation. An indicator needle attached to the shaft of the synchro receiver shows the direction in which the antenna is pointing. In other words, the transmitter converts an angular position into an electrical output, and the receiver converts an electrical input into an angular position. Synchros are not used to turn heavy loads; their use is restricted to position indicators and remotely controlled dials.

A synchro system is diagrammed in Fig. 7-10. In operation, the ac in the transmitter's movable coil induces voltages in the transmitter stator windings. For example, in the position shown, 115 V in the movable coil might induce 50 V in stator winding A (with which the movable coil is aligned) and 25 V each in coils B and C. The connections provided by the control lines will cause these same voltages to appear in the receiver's three stator coils. The receiver stator coils' magnetic fields will interact with the movable coil's field to rotate that coil to the same angular position as the transmitter's movable coil occupies. As you can see, the transmitter and receiver are constructed identically; they are, therefore, interchangeable. In many applications, synchros have been replaced by stepper motor systems.

A device that is similar to a synchro in construction and operation is a resolver. Resolvers have a single primary winding and two secondary wind-

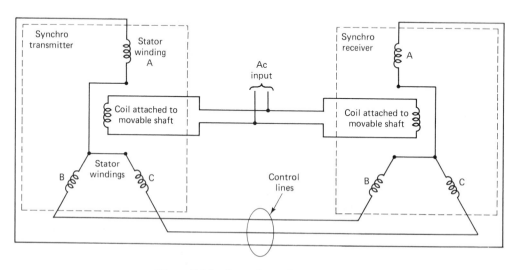

Figure 7-10   Operation of a synchro system.

ings. One secondary produces an output voltage equal to the product of the input voltage and the sine of the shaft angle; the other secondary produces a voltage proportional to the cosine of the shaft angle:

$$E_{sec\ 1} = E_{(in)sin}$$
$$E_{sec\ 2} = E_{(in)cos}$$

Resolvers are used in instruments in which an angular position must be converted into a single voltage level.

SUMMARY

1. Motor-generators are electromechanical devices that convert electrical power from one form (voltage, number of phases, ac or dc) to another form.

2. Dynamotors are smaller, less-efficient devices that do the same job as motor-generators.

3. Stepper motors are devices that rotate a few degrees (7.5°, 15°, or 18° typically) each time the relative magnetic polarity of the two stator windings is reversed. They are rated in terms of step angle, holding torque, residual torque, dynamic torque, operating voltage, and winding type.

4. Gearmotors are motors with built-in gear trains that reduce shaft speed and increase torque.

5. A servo system consists of a motor, a sensing mechanism such as a tachometer, a comparator, and an error amplifier. Servo systems have extremely good speed regulation. Most servomotors are dc PM or ac-single-phase PSC or two-phase. Ac servomotors have a fixed phase winding which is usually 115 V or 230 V and a control winding that may be 15 to 36 V, 115 V, or 165 V.

6. A synchro system is an electromechanical system consisting of a transmitter and a receiver. The shaft of a synchro receiver tracks the angular position of the synchro transmitter shaft.

7. A resolver converts mechanical variations in angular position to variations in the amplitude of an ac voltage.

QUESTIONS

1. Give an application for a motor-generator.
2. What would be three advantages and four disadvantages of using a dynamotor for the application in Question 1?

3. How can a motor-generator's output voltage be varied?

4. Are motor-generators becoming more common or less common? Why?

5. When is an electrical load connected to a motor-generator?

6. What causes RFI to be produced in motor-generators and dynamotors? How can it be eliminated?

7. What does a stepper motor do?

8. What causes a stepper motor to "step"?

9. Name and describe three types of stepper motor.

10. Define:
    (a) Step angle
    (b) Holding torque
    (c) Residual torque
    (d) Dynamic torque
    (e) Operating voltage
    (f) Unipolar (four-coil) winding
    (g) Bipolar (two-coil) winding

11. What waveform does the drive voltage to a stepper motor have? Draw several cycles and indicate the point at which stepping occurs.

12. Describe the operation of:
    (a) L/4R drive
    (b) Bilevel drive
    (c) Chopper drive

13. What are "ringing" and "resonance" in stepper motors?

14. What two things does a gearmotor do that a nongeared motor cannot do as well?

15. What are three disadvantages of a gearmotor?

16. What is a servo system used for?

17. Draw a block diagram of a servo system and describe the function of each block.

18. Name three types of servomotors.

19. Draw a diagram of the proper way to connect an ac servomotor. Indicate the correct voltages to be expected at the motor windings.

20. What does a synchro system do? How many wires are used to connect the transmitter to the receiver?

21. What does a resolver do?

8

TROUBLESHOOTING, REPAIR, AND REPLACEMENT OF MOTORS AND RELATED COMPONENTS

By now you have learned the hows and whys of motor operation, so it's time to start seeing how your knowledge can be put into practice. Troubleshooting of systems containing motors is important for all kinds of systems, ranging from electric hair dryers to large industrial machines. With any type of troubleshooting, the first step is to determine which unit is at fault. Figure 8-1 shows block diagrams of several typical systems using motors and generators. In any of these systems, failure of the motor to operate could be caused by any of a number of problems. The technician's first job is to isolate the faulty item. Just how this is done depends on the type of trouble. If a motor does not work at all, there may be a fault in the circuit that supplies power to the motor. For example, the switch could be defective, or the thermal circuit breaker could be open. Often this kind of initial detective work requires nothing more than a bit of logical thought and a multifunction meter.

Figure 8-2 is a flowchart that describes the basic procedure for troubleshooting a system if the motor is found to be getting the proper power. The next part of this chapter will discuss the principles involved in this flowchart. Then the specific repair procedures will be discussed.

For cases in which the motor is not getting the proper power, the control system feeding the motor must be checked. The last section of the chapter deals with methods of testing control components, including electronic units.

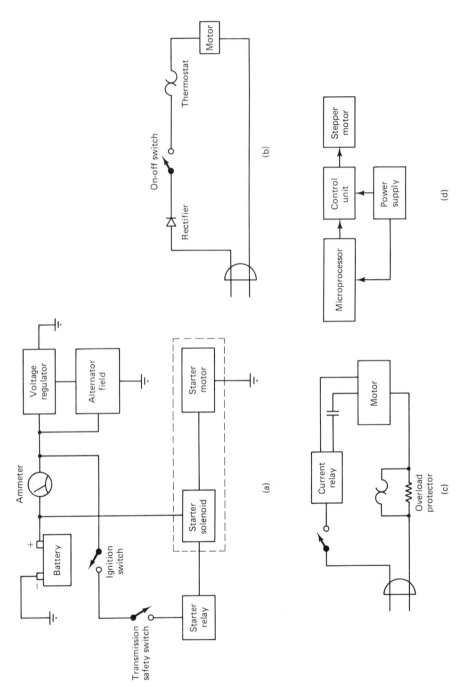

Figure 8-1 Motor and generator systems: (a) automotive charging and starter system; (b) electric hair dryer using PM motor (heater circuit not shown); (c) electric table saw; (d) microprocessor-controlled stepper motor system.

180

THEORY OF MOTOR TROUBLESHOOTING

If a motor is getting power, but is not turning, obviously the motor is at fault. Far too often, the assumption is made that the motor must be rewound or replaced. (Rewinding a motor consists of removing all the old stator and/or rotor winding wire and replacing it with new wire. Since it involves quite a bit of hand labor it is generally economically feasible only for large motors.) Most defective dc or single-phase ac motors can be repaired without rewinding. The first step is to find out whether the problem is mechanical. Hence the first step in the flowchart. Problems with bearings will keep the shaft from turning freely, and may cause audible grinding or scraping sounds. In any event, it will be necessary to disassemble the motor. If the frame is welded together, repair is usually not feasible. Usually, though, motor frames are held together by long screws or bolts. When these are removed, the end bells of the motor housing will usually come off easily. Occasionally, you will find a motor whose housing is press-fitted together. In such cases, a few light taps with a hammer will help to separate the end bells from the housing. Figure 8-3 shows a motor with the end bells removed. If the motor is a PM type, the rotor will cling tightly to the stator once the end bells are no longer there to center it. Being aware of this can prevent pinched fingers! A discussion of motor bearings is given later in this chapter.

If the problem is a foreign object in the motor, this will be immediately obvious once the motor is disassembled. Often, when a foreign object finds its way into a motor, it not only causes the motor to bind but does other damage. Therefore, if a foreign object is found inside the motor case, look for damaged windings, scraped insulation on connecting wire, deformed centrifugal switches, or bent or broken blades on the internal cooling fan.

Gearmotors that have been subjected to sudden impulsive loads while operating can suffer from chipped gears. Inadequate lubrication of a gear train can cause a motor to seize. The problems associated with gear trains are discussed later in this chapter.

If the motor shaft turns freely, the problem is not mechanical. The flowchart lists nine symptoms of possible electrical troubles. We will discuss the causes of these troubles next.

Often, a system containing a motor will develop a defect that is thought to be the fault of the motor, when in fact the problem is a mechanical trouble in the load. To eliminate this possibility, once the motor is determined to be mechanically okay, apply power. If the trouble persists with the load disconnected, the motor is at fault. (*Exception:* Do not apply power to an unloaded series-wound motor, as runaway may result. Series motors having gear trains can be operated without a load, as the gear train itself provides enough frictional load to prevent runaway.)

If the motor does not turn, hum, or get hot when power is applied, check the nameplate to see whether the motor has an internal overload pro-

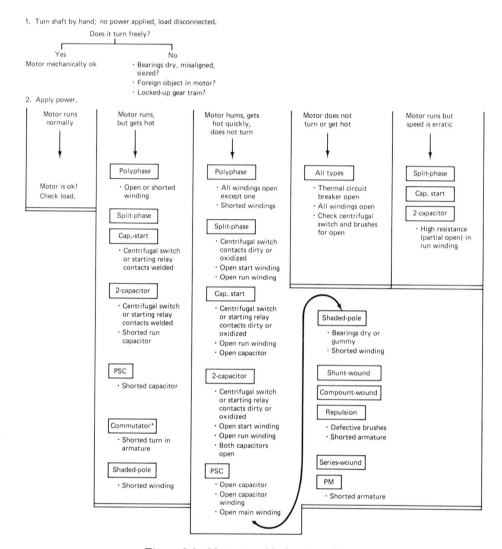

1. Turn shaft by hand; no power applied, load disconnected.

Does it turn freely?

Yes — Motor mechanically ok

No
· Bearings dry, misaligned, siezed?
· Foreign object in motor?
· Locked-up gear train?

2. Apply power.

Motor runs normally	Motor runs, but gets hot	Motor hums, gets hot quickly, does not turn	Motor does not turn or get hot	Motor runs but speed is erratic

Motor is ok!
Check load.

Polyphase
· Open or shorted winding

Split-phase

Cap.-start
· Centrifugal switch or starting relay contacts welded

2-capacitor
· Centrifugal switch or starting relay contacts welded
· Shorted run capacitor

PSC
· Shorted capacitor

Commutator*
· Shorted turn in armature

Shaded-pole
· Shorted winding

Polyphase
· All windings open except one
· Shorted windings

Split-phase
· Centrifugal switch contacts dirty or oxidized
· Open start winding
· Open run winding

Cap. start
· Centrifugal switch or starting relay contacts dirty or oxidized
· Open run winding
· Open capacitor

2-capacitor
· Centrifugal switch or starting relay contacts dirty or oxidized
· Open start winding
· Open run winding
· Both capacitors open

PSC
· Open capacitor
· Open capacitor winding
· Open main winding

All types
· Thermal circuit breaker open
· All windings open
· Check centrifugal switch and brushes for open

Shaded-pole
· Bearings dry or gummy
· Shorted winding

Shunt-wound

Compount-wound

Repulsion
· Defective brushes
· Shorted armature

Series-wound

PM
· Shorted armature

Split-phase

Cap. start

2-capacitor
· High resistance (partial open) in run winding

Figure 8-2   Motor troubleshooting chart.

tector. Sometimes these protectors open and fail to reset. Some overload protectors are easy to replace; others are embedded in the windings and are inaccessible. If you suspect a faulty protector, disassemble the motor and find the protector, then check it with an ohmmeter. When the motor is cool, the protector should measure virtually zero resistance. If it does not, replace it.

Polyphase ac motors are quite simple internally; they have no capacitors, centrifugal switches, brushes, and so on. Therefore, except for faulty

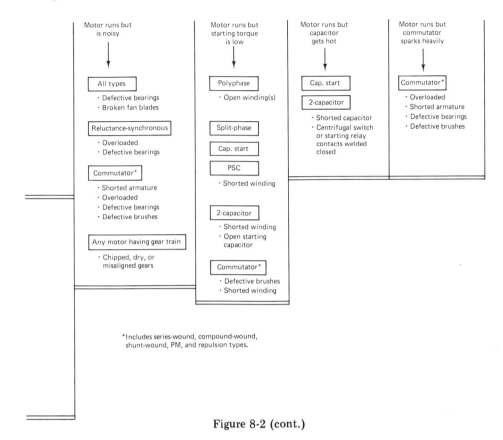

Figure 8-2 (cont.)

overload protectors, almost all nonmechanical problems that occur in a poly-phase motor are the result of open or shorted windings. In fact, it is usually a waste of time even to bother opening up a non-internally-protected poly-phase motor if the shaft turns freely but the unloaded motor does not work properly. Rewinding or replacement are the two choices. The discussion of motor ratings and lead identification later in this chapter may be of help if replacement is necessary.

Single-phase motors having centrifugal switches or current relays are subject to oxidation of switch contacts, especially if the motor is overloaded

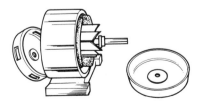

Figure 8-3 Typical motor—disassembled.

or used in applications requiring frequent starting and stopping. When contacts are oxidized, the starting winding will not be energized, but the run winding will. The motor will therefore not turn and no countervoltage will be generated in the run windings, so excessive current will be drawn, causing the motor to heat up rapidly. An open capacitor in a capacitor-start motor will cause the same symptoms, for the same reason: no current in the starting winding.

Centrifugal switches and current relays will also cause trouble if the contacts become welded closed. This can happen as a result of overloading or too-frequent start-stop cycling. When the contacts are permanently closed, current flows in the starting winding continually. Even though the motor will start and run, it will quickly overheat. In a capacitor-type motor, this will usually cause the capacitor to overheat also, and possibly to short and/or explode. (You will remember that starting capacitors are usually ac electrolytics, which have a high ESR and are therefore suitable only for intermittent high-current use.)

Another capacitor problem can occur in two-capacitor motors: a defect in the run capacitor. Since this is usually an oil-bath capacitor, and therefore rather reliable, failure is not very common. However, run capacitors can short or open. If they short, the capacitor winding will draw too much current, and because of the decreased phase difference, the motor's torque will be reduced as well. So low torque and a tendency to run hot could indicate a shorted run capacitor. Low torque alone could indicate an open run capacitor. This would allow the capacitor winding to function almost normally during starting, but that winding would not get any current while running, so that the motor would act as a capacitor-start motor. As you remember, those have less running torque than two-capacitor motors.

Shaded-pole motors have only one common problem. Their starting torque is so low that if the bearings become dry or if the lubrication becomes only slightly gummy, they may not start. The inductance of a typical shaded-pole motor's winding is high enough that inductive reactance usually limits the "stalled-rotor" current to a safe value. Thus, even though shaded-pole motors may heat up, they are seldom destroyed by overheating. (However, it is not unknown for stalled shaded-pole motors to become hot enough to ignite oil that collected on the windings from overzealous lubrication. The resulting fire may destroy a lot more than the motor!)

The various problems that may occur in commutator motors as a result of defective brushes are discussed in detail later in this chapter. Suffice it to say that brushes are the first suspect in many commutator-motor problems.

The final problem to which all motors are susceptible is shorts or opens in the windings. When all other possibilities have been eliminated, windings can be tested as described later in this chapter. If windings are determined to be faulty, there are three options:

1. PM and series-type motor manufacturers often supply replacement armatures. Since the armature is located in a position that makes it difficult for it to get rid of heat, it is more likely to fail than are the field windings.

2. Rewinding.

3. Replacement.

BEARINGS

There are four types of bearings that are commonly used in motors. These are shown in Fig. 8-4. The most common are sleeve bearings. A sleeve bearing is nothing more than a replaceable bushing that surrounds the motor shaft at a stress-carrying point. Sleeve bearings may be made of steel, bronze, oil-impregnated bronze, or bronze with inner recesses filled with graphite (graphitized self-lubricating). The latter two types of bronze bearings are often referred to as "self-lubricating," but in reality all four types of sleeve bearings require lubrication. Usually, motors using sleeve bearings have felt packings on one end of each bearing. The packing is kept saturated with oil, which seeps into the space between the shaft and the inside of the sleeve, providing continuous lubrication.

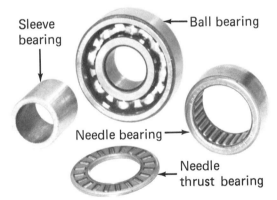

Sleeve bearing — Ball bearing — Needle bearing — Needle thrust bearing

Figure 8-4 Bearings. (Courtesy of Bodine Electric Co.)

Ball bearings are used in more expensive motors because of their greater durability. Also, ball bearings require less maintenance and are less likely to seize up if they are operated without lubrication. However, they do produce more noise than do sleeve bearings. Two lubrication methods are used for ball bearings. Oil-packed ball bearings require periodic lubrication. Grease-packed bearings are often sealed, as they are not expected to require periodic maintenance. They are also less noisy than oil-packed bearings. However, at low temperatures, the grease can become quite stiff. This can make the motor pull up to speed slowly, because of the extra mechanical resistance provided by the bearing. After a few minutes of operation, however, sufficient heat is usually developed to make the bearing turn more freely.

For applications where space is at a premium, but where the lower friction, longer life, and reduced maintenance of ball bearings are desired, needle bearings provide a workable alternative. Although noisier than ball bearings, they provide almost equivalent performance in all other areas. They require lubrication at about the same intervals as do oil-packed ball bearings.

When a shaft is subjected to thrusting pressures (pressures tending to push the shaft lengthwise out of the motor) thrust bearings are required. The most effective of these is the needle thrust bearing. Less expensive (and less durable) alternatives include steel, bronze, and nylon thrust washers. The operation of a thrust bearing or washer is illustrated in Fig. 8-5.

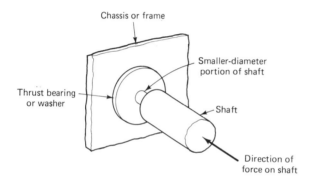

Figure 8-5   Operation of thrust bearings.

As with most things in this world, the type of bearing most likely to give trouble is the cheapest one—the sleeve bearing. However, any type of bearing can give problems. The simplest bearing problem is lack of lubrication. A dry sleeve bearing will make a scraping or sliding sound while turning slowly, and may squeal when turning rapidly. It will also cause the motor to drag, pull up to speed slowly, and perhaps to run somewhat slowly. Lubricating oil is applied directly to the felt wick in some motors, but more often there is a small oil hole in the motor housing or an oil tube protruding through the housing. These last two arrangements are illustrated in Fig. 8-6. In any case, the oil should be added only until the wick is saturated—not dripping. Usually after a bearing has been allowed to run dry, it takes some time before the new oil works its way into all the proper places, so the motor may continue to make noise for perhaps a minute or two when power is applied again. As for the oil itself, 10-weight nondetergent motor oil is the best, although in a pinch 20- or 30-weight oil can be used. So-called "machine oil" that is sold in cute little squeeze cans is usually not suitable for electric motors, because under continuous use the bearing may become hot

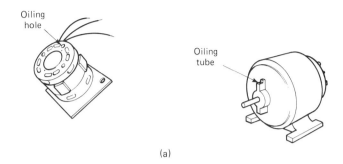

(a)

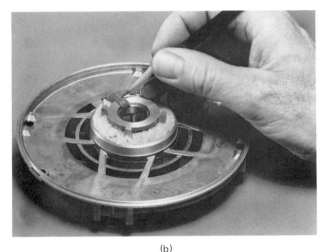

(b)

Figure 8-6 Lubrication of sleeve
bearings: (a) oiling arrangements;
(b) detail of lubricating wick
(courtesy of General Electric Co.)

enough to change that kind of oil into a gummy, varnish-like substance. Detergent-type oils are not acceptable, because the detergents can attack the insulation of the wire with which the motor is wound.

One final important comment about oiling motors: too much oil—or oil in the wrong places—is almost as bad as no oil. Getting oil all over the housing of any motor will interfere with the motor's ability to dissipate heat. The best approach to motor lubrication is regular maintenance—usually a few drops of oil for each bearing every 3 months or so. But even if a bearing dries out completely, do not respond by oiling the bearing until it drips!

The combinations of oil and dust, inferior oil and heat, or any oil and long years of service tend to produce gum. A common bearing problem is the increased resistance to rotation caused by such gum. When a bearing has become gummy, the remedy is to wash the bearing and shaft thoroughly with Varsol or lacquer thinner, then relubricate. When using these solvents, be very careful to keep them off the motor windings, as they can attack the insulation.

If the shaft can be wiggled in the bearing, the bearing should be replaced. This wiggling does not apply to end-to-end sliding of the shaft, but rather to (1) excessive clearance (or "play") between the shaft and the inside circumference of a sleeve bearing, or (2) worn balls or ball races in a ball bearing. Ball bearings can often be removed by gently tapping the end bell in which the bearing is mounted, with the bell placed so that gravity pulls out the bearing. Sleeve bearings are usually press-fitted, and can be removed by use of an arbor press. If no press is available, a cylindrical object and a hammer can be used to gradually force the old bearing out and then to press the new one in (see Fig. 8-7). When installing a new bearing, be careful to align the oil holes (if any) with the holes in the motor housing. Also, take care not to burr or deform the new bearing. Sometimes new bearings are supplied slightly undersize, and a reamer must be used to size the bearing appropriately so that the shaft will turn freely in it.

Figure 8-7   Removal of sleeve bearing.

    If a sleeve bearing is operated for a long period of time without lubrication, it may become so hot that the inner surface will actually melt and fuse to the shaft. The bearing is then referred to as either "welded" or "frozen." Frozen bearings must first be removed from the shaft. If you are lucky, a hammer can be used to break it loose. In extreme cases, the bearing may have to be removed with a torch. Often, a propane torch can be used rather than an oxyacetylene one, reducing the chances of damaging the shaft. In either case, once the bearing is off, the shaft may be scarred, pitted, or grooved. Any lumps on the surface should be carefully filed away. Surface scratches should then be polished out with a fine grade of emery paper. Deep scratches, pits, or grooves should be built up, either by welding or by use of a steel-filled epoxy. The shaft must then be turned back down to the proper size in a lathe. It is very important to avoid bending the shaft during this procedure. Complicated as all this sounds, a damaged shaft can usually be reclaimed in less than an hour's working time, provided that the necessary equipment is available. This is certainly less expensive than replacing the motor.

In reassembling a motor with sleeve bearings, it is easy to get the end bells tightened unevenly, so that the bearings are misaligned. The result is that the shaft will be difficult to turn or will not turn at all when the motor is completely reassembled. To avoid this, the end bells should be tapped carefully onto the main motor housing until the hammer makes a solid sound all the way around. Then the screws should be tightened gradually. The correct 1-3-2-4 sequence for tapping on the end bell and for tightening the screws is shown in Fig. 8-8. Each screw should be tightened about ½ turn at a time, until all screws are tight.

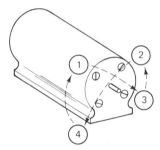

Figure 8-8   End-bell tightening sequence.

GEAR TRAINS

Large gearmotors designed for maximum life use oil to lubricate the gear train, and some method for periodically checking and replenishing the oil supply is provided. This information should be found on the nameplate of the motor or gearbox. Most small gearmotors use grease rather than oil for a lubricant. Whenever a gear train begins to become noisier than usual, the grease supply should be checked. It can be replenished with a molybdenum disulfide ("moly") grease or any other good-quality gear lubricant.

STARTING SWITCHES

Figure 8-9 shows a variety of types of starting switches used on split-phase, capacitor-start, and two-capacitor motors. The most common defect for these switches is oxidized or dirty contacts. Contacts can be restored by the same procedures used for relay contacts (see Chapter 3). The other defect to which the mechanical switches are susceptible is welding. If contacts are welded together, the entire starting switch or relay must usually be replaced. Occasionally, though, the contacts can be separated and re-formed sufficiently well to avoid the need for replacement. The procedure for re-forming

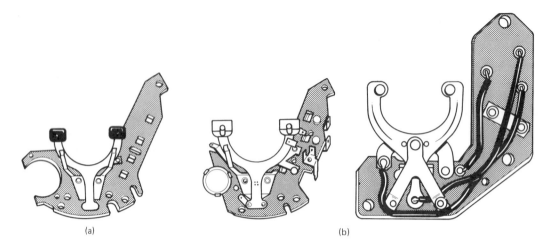

(a)   (b)

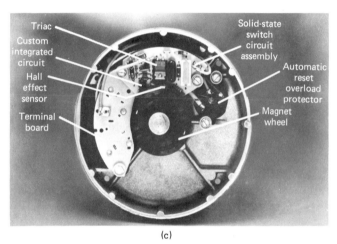

(c)

Custom
integrated
circuit

Triac

Hall
effect
sensor

Terminal
board

Solid-state
switch
circuit
assembly

Automatic
reset
overload
protector

Magnet
wheel

Figure 8-9 Starting switches:
(a) single-contact; (b) three-contact;
(c) solid-state starting switch:
single-speed design (courtesy of
Century Electric, Inc.).

contacts is given in Chapter 3. Whenever you are working with switch contacts, be very careful not to deform the springs, or the switch may not operate properly even with clean contacts.

Solid-state starting switches will probably gradually replace mechanical ones. However, they are also subject to failure, usually in the form of a shorted triac. Testing electronic components is discussed later in this chapter.

MOTOR CAPACITORS

Capacitors usually die either by leaking, shorting, or opening. They can be tested by means of an ohmmeter. Set the ohmmeter to the highest resistance scale. Then touch the probes to the two capacitor leads. The needle should kick very briefly, indicating a low resistance; then the resistance reading should increase—slowly for electrolytics, more rapidly for oil-bath capacitors. The larger the capacitor, the slower the rate of increase. Finally, the reading should level out at some high value. An electrolytic should end up reading 100 kΩ or more; and an oil-bath capacitor should read almost infinity (like an open circuit).

If the meter needle does not kick at all when the probes are first connected, briefly short the capacitor leads together and try again. If the needle still does not move, the capacitor is open. If the needle does not level off at a high enough resistance, the capacitor is leaky or shorted.

Open, leaky, or shorted capacitors must be replaced with units having the same capacitance and the same or higher voltage rating. Always replace an oil-bath capacitor with an oil-bath, never with an electrolytic. However, an electrolytic can be replaced with an oil-bath.

The testing method just described does not work well with capacitors below 1 μF, but virtually all motor capacitors are larger than 1 μF. Occasionally, you may find a capacitor that leaks or shorts when full voltage is applied but checks okay with an ohmmeter. In this case, the capacitor can be tested on a unit that applies the full rated voltage to the capacitor, or it can be tested by substitution. If you substitute a new capacitor and it corrects the problem, the old one was bad.

BRUSHES

Brushes for electric motors may be made of carbon, synthetic graphite, natural graphite, or a metal-graphite composite (see Fig. 8-10). The manufacturer chooses one of these materials over the others because of current-carrying capacity, wear characteristics, and maximum allowable commutator speed. Brushes are held against the commutator by means of a spring. The spring itself may be a coil spring (ballpoint-pen type), or a roll spring (watchspring type).

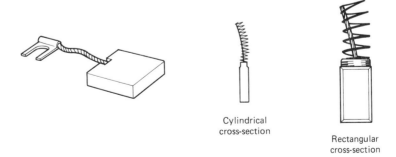

Cylindrical
cross-section

Rectangular
cross-section

Figure 8-10   Typical brush assemblies.

Brushes can fail for several reasons. The most common is wear. A brush is worn out if there is so little of it left that it cannot be held securely by the brush holder. Generally, this means a minimum length of $\frac{1}{4}$ in. A brush that has been subjected to too much current, either because of the motor being overloaded or because of a defective armature, can undergo a change in the structure of the brush material that makes it a poor conductor. Such brushes have a blackened, crystallized appearance, are often pitted, and are quite brittle. These must be replaced. The third cause of brush failure is oil. If a motor is used in a location where oil spray is common, or if the motor is improperly lubricated, the commutator surface can become covered with oil. This insulating film interferes with the current on its way to and from the armature. Cleaning the armature with denatured alcohol will remove the oil and may restore proper operation. However, if the brushes themselves— which are microscopically porous—have become soaked with oil, they will continuously reoil the commutator. Therefore, oil-soaked brushes must be replaced. A word of caution is in order, though. Graphite and metal-graphite brushes are somewhat slippery, so it would be easy to suspect them of being oil-soaked when they are not. As a general rule, an oily residue will appear on the surface of a brush that is truly oil-soaked.

Some technicians make the mistake of replacing brushes whenever there is a lot of sparking during operation. While sparking is often a sign of defective brushes, it can also indicate several other problems:

1. Armature drawing too much current
2. Faulty brush holders
3. Worn commutator
4. Weak or broken brush springs

The armature will draw excessive current if the motor is overloaded either externally or as a result of bearing or gear-train defects. A shorted winding will also result in excessive armature current.

Brush holders are often made of some form of plastic. Age, exposure to heat, and chemical fumes can cause plastic to deteriorate. If a brush holder becomes incapable of holding the brush firmly (i.e., if the brush has a lot of "free play"), the holder needs replacing.

Although brushes do by far most of the wearing, the commutator itself will wear somewhat also. Therefore, an old motor that has gone through several sets of brushes may require resurfacing of the commutator. If the commutator has only minor surface scratches, these can be removed by hand, using No. 0000 emery paper. However, if the armature is badly worn or scarred, it must be machined. This is done by clamping the armature in a lathe and removing a very small amount of commutator material—just enough to make the commutator smooth and perfectly cylindrical. After a commutator is resurfaced, the insulating material between commutator segments may need to be undercut, which is a process requiring a special piece of equipment. If the level of the insulator is below that of the commutator segments, as shown in Fig. 8-11, undercutting is not necessary.

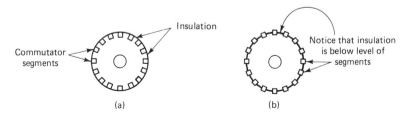

Figure 8-11   Undercutting of commutator: (a) commutator that needs undercutting; (b) properly undercut commutator.

Obviously, when brush springs are broken, they must be replaced. Sometimes, though, coil-type brush springs weaken so that they do not provide enough pressure on the brush. Such springs should be replaced, but an emergency procedure is to stretch the spring slightly and reinstall it. How much brush pressure is enough? Well, it is largely a matter of judgment. The pressure must be sufficient to keep the brush firmly seated against the commutator as it turns, but not so much as to cause unnecessary wear.

When installing brushes, either new or old, be careful to orient them correctly. Figure 8-12 shows proper and improper installation. Notice that the properly installed brushes fit closely against the commutator.

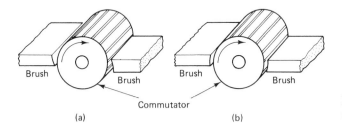

Figure 8-12   Brush installation; (a) improper; (b) proper.

THERMAL PROTECTION

Since the most expensive repair that a motor is likely to need is rewinding, and winding defects usually result from overheating, manufacturers often incorporate devices to interrupt the current to a motor's windings in the event of overheating. There are two families of these devices. The most common is simply a temperature-sensitive switch that is mounted on the motor case or in the winding. Such switches are typically constructed with one contact mounted on a bimetallic disk that, when heated, changes suddenly from a concave to a convex shape, breaking the circuit to the motor windings. Figure 8-13 shows both case-mount and in-the-winding thermal protectors.

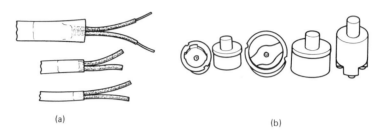

(a)

(b)

Figure 8-13   Thermal protectors. (Courtesy of Bodine Electric Co.)

The second family of thermal protectors is sensitive to both temperature and current. They are constructed much like the first type, with the addition of a small heating coil in proximity to the bimetallic disk. When the motor draws enough current for a long enough time to heat up the disk so that it snaps into its "open" configuration, the circuit is interrupted. This second type of protector has two advantages. The first is that it can be mounted in any convenient location, not necessarily on the motor. This makes this type of protector much easier for the operator to get to. Second, current-sensitive protectors can be made to open more reliably when a potentially damaging situation exists, because they are not slowed down by thermal time lag—the time required for the heat of an overheating winding to reach the protector's bimetallic disk.

Either type of protector is available in either self-resetting or manually resetting versions. The former version simply remakes the connection to the windings whenever the winding temperature drops sufficiently. The second type must be reset by hand after the temperature has dropped to a safe level.

Motors that incorporate thermal protection usually include the words "thermally protected" or "overload protected" somewhere on the motor case. Whenever a thermally protected motor does not run, the problem is probably an open protector. The technician should make sure that the motor has had ample time to cool down. (This may take 30 minutes or more in some cases.) Then the reset switch should be pressed, if there is one. If the motor still will not restart, there is a very good chance that the thermal protector itself is defective. Some such protectors last for many cycles of operation, but others seem to die on their first operation. Since a thermal protector is just a heat-operated switch, it is tested by checking for continuity with an ohmmeter. A very low reading—a fraction of an ohm—should be obtained.

Oh, yes . . . in-the-winding protectors can be rather difficult to locate. It should go without saying that finding one requires opening up the motor. It may also require carefully cutting some of the tape or lacing that binds the windings. Once a defective thermal protector has been located, it should be replaced only with a new one of the same type.

WINDINGS

Special equipment is available for detecting shorted windings. However, the average technician does not have access to such equipment, and must get along with the trusty VOM. Two tests can be made with a VOM to show up defective windings. The easiest, of course, is simply to check the continuity of the winding: Does it conduct current? An open winding will indicate infinite resistance, whereas a good one will indicate a low resistance. Remember the schematic diagram of the motor you are testing, though (Chapters 4 to 7). It is possible to make a too-hasty diagnosis of an open winding when the problem is really something else open that is in series with the winding, such as thermal protectors, switches, capacitors (should be open to dc!), or brushes.

The second VOM test for a defective winding is to measure from one winding lead to the motor case. For a motor having a wound rotor, this test is made by measuring from each commutator segment or slip ring to the iron commutator core. In either case, the VOM should indicate an open circuit. Any other reading indicates a defect. One caution, though. Motors and generators used in autos, tractors, and so on, often have one lead of the stator winding deliberately connected to the case. Before testing such motors for a short to case, be sure that any such leads are disconnected.

This brings us to the subject of identifying motor leads. Sometimes a motor nameplate tells which color wire goes to which winding. With a wound-armature motor, identifying the field and armature leads is easy. Single-speed, unidirectional capacitor-start, PSC, and two-capacitor motors also present no problem: the leads that go directly to the power line are the main winding leads, and the other leads go to the capacitor winding. But split-phase, polyphase, dual-voltage (115/230, or 220/440 V), and multi-speed motors can be more of a challenge. There is a general rule for single-phase motors: the starting or capacitor winding always has a higher resistance than the main winding. If the motor is a multispeed or dual-voltage type,

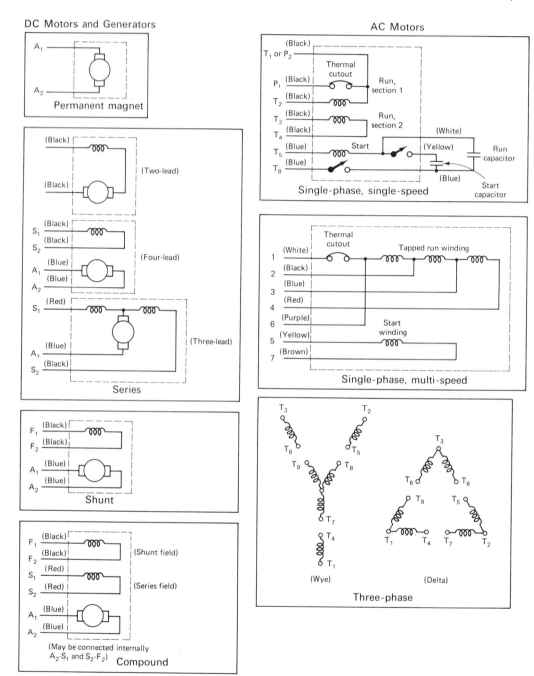

Figure 8-14 Color codes and terminal identification for motor leads.

there may be several sections to the main winding, or even several separate "main" windings.

Aside from the foregoing rules, you would be on your own except that most manufacturers now adhere to a standard color code and terminal designation system for motor leads. These are shown, for various types of motors, in Fig. 8-14. In using this figure to determine the proper connection for a specific motor, you must use several things you have already learned. Specifically:

1. Since series, shunt, and compound motors are reversed by reversing the polarity of the field leads with respect to the armature, the external connections of the terminals depend on the direction of rotation desired.

2. The shunt field (F terminals) is always connected in parallel with the armature; the series field (S terminals), in series with it.

3. The direction of rotation of a single-phase ac motor depends on the direction of current in the starting winding (T_5 and T_8) compared to that in the main winding (other T terminals).

4. Although series motors are shown under the heading "DC Motors and Generators," they can be used with dc or ac.

5. The starting switch of a single-phase motor may take the form of an external potential relay or current relay.

Also, you must understand that these are very general diagrams, so that specific motors may not have all the connections shown. For example, a split-phase motor will have no capacitor leads. A capacitor-start motor will have no run capacitor leads. A non-thermally-protected motor will have no protector (P) leads; the run windings will terminate in T_1-T_2 and T_3-T_4. A PSC motor will have no internal starting switch, nor will a capacitor motor using an external current relay or potential relay. A single-voltage motor will have only one run winding, connected to T_1 and T_4. Nonreversible single-phase motors may have the start and run winding interconnected internally.

Finally, you need to know how dual-voltage motors are connected. For example, a 115/230-V motor will have two run windings and a start winding. For low-voltage (115-V) operation, all three windings are connected in parallel. (Naturally, the start winding will have its switch and any capacitors in series!) For high-voltage operation, the run windings are connected in series, with the start winding in parallel with one of them.

With all these things in mind, you can understand the connections specified in Table 8-1. In the table you see references to L connections. These are power-line connections. Thus a single-phase motor is powered by L_1 and L_2; and a three-phase motor by L_1, L_2, and L_3.

TABLE 8-1 Motor Connections

Type of motor	Clockwise rotation[a]	Counterclockwise rotation
PM motor	$+$ to A_1, $-$ to A_2	$+$ to A_2, $-$ to A_1
Four-lead series	L_1 (or $+$ or $-$) to A_1, A_2 to S_1, L_2 to S_2	L_1 (or $+$ or $-$) to A_1, A_2 to S_2, L_2 to S_1
Three-lead series	L_1 (or $+$ or $-$) to A_1, L_2 to S_1	L_1 (or $+$ or $-$) to A_1, L_2 to S_2
Compound motor	L_1 (or $+$ or $-$) to A_1 and F_1, A_2 to S_1, L_2 to S_2 and F_2	L_1 (or $+$ or $-$) to A_2 and F_1, A_1 to S_1, L_2 to S_2 and F_2
Nonprotected single-phase		
Single voltage		
Dual voltage—low	L_1 to T_1 and T_5, L_2 to T_4 and T_8 L_1 to T_1, T_3, and T_5; L_2 to T_2, T_4, and T_8	L_1 to T_1 and T_8, L_2 to T_4 and T_5 L_1 to T_1, T_3, and T_8; L_2 to T_2, T_4, and T_5
Dual voltage—high	L_1 to T_1; T_2 and T_3 to T_5; L_2 to T_4 and T_8	L_1 to T_1; T_2 and T_3 to T_8; L_2 to T_4 and T_5
Three-lead reversible capacitor[b]	L_1 to T_1 and capacitor lead; L_2 to T_4	L_1 to T_1; L_2 to T_4 and capacitor lead
Protected single-phase[c]		
Single voltage		
Dual voltage—low		
Group I	L_1 to P_1, T_1 to T_5, L_2 to T_8	L_1 to P_1, T_1 to T_8, L_2 to T_5
Group II	L_1 to P_1; P_2 and T_3 to T_5; L_2 to T_2, T_4, and T_8	L_1 to P_1; P_2 and T_3 to T_8; L_2 to T_2, T_4, and T_5
Group III	L_1 to P_1; P_2 to T_3; T_1 to T_5; L_2 to T_2, T_4, and T_8	L_1 to P_1; P_2 to T_3; T_1 to T_8; L_2 to T_2, T_4, and T_5
Dual voltage—high		
Groups I and II	L_1 to P_1; P_2 to T_5; T_2 and T_3 to T_8; L_2 to T_4	L_1 to P_1; P_2 to T_8; T_2 and T_3 to T_5; L_2 to T_4
Group III	L_1 to P_1; T_2 and T_3 to T_5; L_2 to T_4 and T_8	L_1 to P_1; T_2 and T_3 to T_8; L_2 to T_4 and T_5

Multispeed single-phase[d]

Three-speed nonprotected, nonreversible

L_1 to white
L_2 switched to:
   Black for high speed
   Blue for medium speed
   Red for low speed
   Yellow through external
      capacitor to L_1

or

L_1 to white
L_2 to yellow and switched to:
   Black for high speed
   Blue for medium speed
   Red for low speed

Three-speed protected, nonreversible

L_1 to white; external capacitor from yellow to purple; L_2 connected to brown and switched to black, blue, or red

Three-speed protected, reversible

L_1 to white; L_2 switched to black, blue, or red; yellow, brown, and purple as pictured:

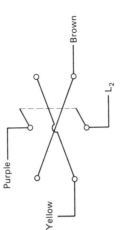

Purple
Yellow
Brown
L_2

Three-phase dual-voltage[e]

Low voltage—wye	L_1 to T_1 and T_7; L_2 to T_2 and T_8; L_3 to T_3 and T_9; T_4 and T_5 to T_6
Low voltage—delta	L_1 to T_1, T_6, and T_7; L_2 to T_2, T_4, and T_8; L_3 to T_3, T_5, and T_9
High voltage—wye	L_1 to T_1; L_2 to T_2; L_3 to T_3; T_4 to T_7; T_5 to T_8; T_6 to T_9
High voltage—delta	L_1 to T_1; L_2 to T_2; L_3 to T_3; T_4 to T_7; T_5 to T_8; T_6 to T_9

[a] Rotation is specified for dc motors from a vantage point facing the commutator end of the shaft. For ac motors, the vantage point is facing the end of the shaft away from the load.

[b] Three-lead reversible capacitor motors have two-section run windings with T_2, T_3, and T_8 internally connected together.

[c] There are three basic methods for connecting protected motors; the connection patterns are therefore designated as group I, II, or III.

[d] Two-speed motors have no red lead.

[e] To reverse the direction of rotation of a three-phase motor, just interchange any two of the power connections (e.g., L_1 and L_2, or L_1 and L_3, etc.).

RATINGS OF MOTORS

When all else has failed and a motor must be replaced, you will need to know certain things about the motor you are replacing. These are summarized below.

1. Type of motor (series, shunt, split-phase, capacitor-start, etc.)
2. Horsepower
3. Operating voltage
4. Number of speeds
5. Type of bearings
6. Type and size of shaft
   a. Diameter
   b. Length
   c. Protrudes from one end or both ends of motor?
   d. Round, flatted, or keyed? (see Fig. 8-15)

Round   Flatted   Keyed   **Figure 8-15** Motor shaft styles.

7. Type of enclosure
   a. Open-frame (you can see the inside)
   b. Totally enclosed (for dusty locations)
   c. Explosion-proof (for locations where there are explosives, fumes, or vapors)
8. Frame size (the National Electrical Manufacturers' Association, NEMA, has established standard sizing for motor frames so that if the frame size of a replacement motor matches that of the original, the replacement will fit physically)
9. Type of mounting (you may have to describe it)

Much of this information can be found on the motor's nameplate. Specifically, items 2, 3, 4, 8, and perhaps 1 should always be listed on the nameplate. In addition, the nameplate should give the manufacturer's type number, which covers all the other necessary information, provided that the necessary catalogs or cross-reference sheets can be located. In the absence of such sheets, you may have to supply the remaining information from your examination of the motor.

A note is in order here about power—how much a motor uses and how much it produces. Electrical power can be put to work in many different ways: to produce heat, to turn motors, to produce sound or light . . . and there are many other uses. Most often, not all of the electrical power that is fed to a device is converted into the desired form of energy. For example, in a motor, some of the energy is lost as heat, because of the resistance of the wire in the motor. In a light bulb, some of the electrical power (watts) is turned into heat rather than light. Usually, we do not think of mechanical power in terms of watts, but rather in terms of horsepower. One horsepower is equal to 746 W.

How well a device changes energy from one form into another desired form is its *efficiency*. Thus a 90% efficient motor turns 90% of the input electrical power into rotational mechanical power at the output. A 90% efficient light bulb changes 90% of its input electrical power into light. The other 10% in each case is lost as other forms of energy, such as heat or sound—primarily heat. Thus the equation for efficiency of any device is

$$\eta = \frac{\text{output power}}{\text{input power}} \times 100\%$$

where η (Greek lowercase letter eta) represents the efficiency. A motor that draws 4 A at 110 V (4 $\times$ 110 = 440 W) and produces an output of 1/2 hp (373 mechanical watts) would have an efficiency of 373 W/440 W $\times$ 100% = 84.7%.

Other information is also included on a motor's nameplate, some of which would be difficult to determine otherwise:

1. Duty cycle (or time rating)—continuous or intermittent.

2. The words "thermally protected," if applicable.

3. Full-load operating current.

4. A code letter, indicating the locked-rotor kVA per horsepower. (This figure is of use to electricians who must wire circuits to feed motors. They have tables in which they can look up the code letter.) The code letter also classifies the motor in terms of starting current and starting torque.

5. A design letter, indicating the range of slip for which the motor is designed. The four most common are:

    A: high torque, low slip, high starting current, highest efficiency.

    B: normal torque, normal slip, normal starting current, less efficient than A.

    C: high torque, normal slip, normal starting current, less efficient than B.

    D: high starting torque, high slip, less efficient than C.

6. Temperature rise. This is the maximum number of degrees Celsius that the motor's temperature will rise above the temperature of the room in which it is used.

7. Service factor. This is the maximum overloading that may safely be applied to the motor. For example, a 2-hp motor with a service factor of 1.15 could be safely required to deliver $2 \times 1.15 = 2.30$ hp, without overheating.

8. Form factor. This rating—for dc motors only—specifies how pure the dc must be in order for the motor to run properly. Pure dc has a form factor of 1. Unfiltered full-wave-rectified dc has a form factor of 1.11. Mathematically, the form factor is the ratio of rms current to average current. If a motor is used with dc having too much ripple (too high a form factor), overheating may occur.

9. Efficiency index. Indicates minimum efficiency and typical efficiency of the motor. Recently introduced, this rating is becoming more common.

10. Torque. The number of ounce-inches, foot-pounds, and so on, of torque that the motor can produce. Listed only for control and torque motors.

11. Inertia. Specified only for control motors, this is the mechanical inertia of the rotating parts of the motor.

NOISE AND VIBRATION CONSIDERATIONS

It is not within the purpose of this book to describe engineering methods for reducing the noise produced by normally operating motors. However, there are several conditions that a technician should be aware of that can cause a motor to be too noisy when it seems to be operating normally in other respects.

The normal noise produced by a brush-type motor varies from a "frying" sound to a whine, depending on the speed of operation. If this sound seems to be accompanied by a crackling, rattling sound or by extraordinary noise, look at the brushes to see if there seems to be undue sparking. Admittedly, deciding how much is too much requires a certain amount of judgment that will improve with experience. However, with good brushes and holders, excessive sparking usually indicates either that the motor is operating at too low a speed (due to overload or defective bearings or gear train) or that a shorted armature is causing the brushes to carry excessive current.

The normal noise produced by induction motors is a low hum. (The exception is reluctance-synchronous motors, which can make a slightly clattering sound.) All the following noise-producing troubles that affect induction motors can also affect brush-type motors.

If a motor seems to have quite a bit of vibration, perhaps accompanied by a low growling noise, either a loose sleeve bearing, a worn ball bearing, or dynamic unbalance may be the problem. Dynamic unbalance is the gremlin that is removed from your car when you have your wheels balanced. It is often caused by broken blades on the motor's internal cooling fan, or by an accumulation of dirt on the fan blades. Broken fans can be replaced; dirty ones can be cleaned. Checking worn bearings has already been discussed. If there is some question about whether a certain amount of noise or vibration is normal, because of the electromechanical operation of a motor, check the motor as follows: Apply power and bring the motor up to speed; then remove power, but do not apply either mechanical or electrodynamic braking. If the noise remains as the motor coasts to a stop, it is likely to be either bearing noise or dynamic unbalance.

Seriously worn or dry ball bearings will produce a raspy scraping or squealing sound. Seriously worn sleeve bearings will produce a pounding noise.

Dry gear trains will make a loud clattering noise. Even properly lubricated gear trains will make some clattering noise, especially if lightly loaded. This is called mechanical backlash noise, and is normal and not easy to remedy.

Several simple methods for limiting the amount of noise radiated into a room by a motor include resilient mountings, proper choice of mounting surfaces, and enclosure of the motor. Resilient mountings are useful for reducing vibration and very low frequency rumble. (In the business, we call this "structure-borne noise.") In order to be effective, the mountings should "give" appreciably when the weight of the motor is applied. Springs, for example, should "give" about half of their maximum travel. In all but the simplest cases, an engineer trained in shock and vibration isolation should be consulted regarding a choice of resilient mountings. Incidentally, where the motor shaft is rigidly coupled to the load so that alignment is critical, the load and motor must be mounted rigidly to a single frame member, and the entire frame member can then be resiliently mounted. In belt-drive systems, the motor alone can be resiliently mounted.

It seems obvious, but is often overlooked, that mounting a motor on a thin panel or piece of sheet metal is asking for acoustical amplification of noise, especially in the middle-frequency range. Wherever possible, motors should be mounted to firm, acoustically dead panels, such as concrete, cast metal, or thick wood or plywood. If thin panels must be used, the noise can be somewhat reduced by bracing and damping the panels. The most common damping method is to apply a heavy, viscous material called "lagging" to the panel. Automotive undercoating is a common example. Lagging materials designed for industrial applications are available at industrial supply houses that handle chemicals and coatings.

High-frequency noise, such as that produced by brushes or high-speed fans and ball bearings, is most easily reduced by enclosing the motor in a separate housing. Such a housing should be made of acoustically dead materials, *not* of sheet metal. In severe cases, the housing can include layers of heavy material such as lead, or can be brushed with a lead-loaded lagging on the inside. Sometimes, acoustical absorbing material such as a 2-in.-thick glass fiber blanket is needed inside the enclosure. Whenever a motor is enclosed to limit noise, adequate ventilation must be provided for the motor. These ventilation ports themselves can allow a significant amount of noise leakage. Once again, except for simple cases, a qualified engineer should be consulted before a motor enclosure is built.

TESTING OF RELATED CIRCUITS AND COMPONENTS

Components

If a fault in a motor system's operation is a result of problems in a speed-control unit, power supply, or other electronic subsystem, the first step in troubleshooting that subsystem is visual inspection. Swelled components, resistors with bubbles on the surface, or capacitors with whitish deposits (dried electrolyte) nearby are suspect. Cinders and ashes are a dead give-away. If there is any doubt about a component, test it with an ohmmeter as follows:

Resistors.   The ohmmeter should register within the rated tolerance of the resistor.

Capacitors.   Check for shorts or opens; capacitors rarely change values. Large ones should kick the ohmmeter needle over to a low resistance, then climb back up to some high value. For large electrolytic capacitors the reading will climb slowly; for smaller capacitors, the climb will be more rapid. Capacitors smaller than about 0.5 μF may not kick the needle perceptibly. A final reading less than 100 kΩ for an electrolytic or less than 5 MΩ for a nonelectrolytic indicates a leaky capacitor that should be replaced. Use of a capacitor checker that applies the full rated voltage is more accurate, if such equipment is available.

Diodes (Rectifiers).   Use the 1-kΩ scale. The ohmmeter should measure low resistance with the $-$ probe on the cathode and the $+$ probe on the anode; should read open-circuit (infinite resistance) with the probes reversed.

Transistors.   Use the 1-kΩ scale. Bipolar (i.e., NPN or PNP) transistors should measure as two diodes connected cathode to cathode (PNP) or anode to anode (NPN). Connected $+$ probe to base, an NPN should measure low resistance with the $-$ probe on emitter or collector. With the $-$ probe on the base, the NPN should measure as an open circuit to emitter and to collector. The PNP should measure exactly opposite; the $-$ probe on the base should give low resistance, and the + probe should give infinite resistance. The final check is for emitter-to-collector leakage. Connect the + probe to the collector of the NPN (− probe for the PNP) and measure to the emitter. It should read infinite resistance.

Power FETs are much more difficult to test than are bipolar transistors. In fact, it is usually better to try to use the process of elimination than to attempt to test these devices. If everything else in the circuit is okay, replacement of the power FET is the next logical step. First, though, a caution. Power FETs can easily be destroyed by discharges of static electricity, so they require special handling precautions. Whenever a power FET is not actually installed in a circuit or being tested, the leads should be shorted together. When one of these devices is being soldered into a circuit or desoldered from a circuit, a grounded-tip soldering iron should be used.

SCRs. Connect the + probe to the anode and the − probe to the cathode. It should read infinite resistance. While holding these connections, briefly connect the gate to the anode and release. The reading should drop to low resistance and "latch" there. (Sometimes a scale other than 1 kΩ may have to be used to make an SCR latch.)

Triacs. Test just as for SCR, then reverse the probes and retest (− probe to anode 2, + probe to anode 1; short gate to anode 2). Should give roughly the same readings for both polarities.

Switches. The ohmmeter should read very low resistance when contacts are "made"; infinite when contacts are open.

Rheostats or Potentiometers. The reading should change smoothly from zero to the rated value as the shaft is rotated. Replace if the reading is erratic or if there are dead spots (infinite resistance).

Transformers, Chokes, and Coils. Each winding should measure less than 1 kΩ. Windings should show infinite resistance to any other winding and to the core or case.

Circuit Breakers. Should read very low resistance; if not, press the reset button and retest.

Motor Starters. Should read progressively lower resistance as shaft is rotated. Watch for dead spots that may indicate defective contact points or burned-out resistor segments.

All these components should have at least one lead disconnected from the circuit when measurements are being made. Otherwise, other components in the circuit may be connected in parallel across the device you are testing, giving erroneous readings.

Speed Control Units

Two specific kinds of electronic subassemblies that you are likely to encounter often are speed control units and power supplies. Speed control units most often fail because of a defective triac or SCR. If this unit checks okay, test the other components one at a time. Sometimes a triac with an internal diac is used; these can only be checked for short circuits via an ohmmeter, because 25 to 35 V is needed to trigger the diac. Similarly, an external diac cannot be tested with an ohmmeter except to see whether it is short-circuited. If the triacs and diacs are not shorted, but the circuit does not work, check all other components. The triac or diac may be selected as the culprit by process of elimination.

Question: How do you know whether a triac has an open gate connection or has an integral diac? *Answer:* Look it up. Several semiconductor manufacturers produce lines of replacement units and guides for selecting which such unit replaces a specific original part. Look up the part you are testing, then find the specifications of the listed replacement unit (usually in the front of the guide). These will tell you whether the triac has an integral diac. For that matter, the same method can be used to identify whether a certain hunk of plastic and metal is a transistor, an SCR, or a triac; the three look the same externally. Also, they will help you to choose replacement units—but more on this later.

Power Supplies

Repairing a power supply begins with the usual visual inspection. If correction of obvious problems does not bring the supply back into proper operation, the next step is to isolate the part of the unit that is most likely to be causing the problem. A review of the block diagrams of Fig. 1-28 may be of help. Trouble isolation begins with an examination of the complaint. Usually, there are three possibilities: It blows fuses, the output voltage is wrong, or there is no output voltage. Let's look at these one at a time.

Blown Fuses. Since the purpose of a fuse is to protect components and wiring against excessive current, a blown fuse (or a circuit breaker that keeps popping out) indicates that something is trying to draw too much current. Often, the problem is in the load: that is, whatever the power supply is supplying current *for.* So the first step is to disconnect the load. If fuses or breakers still pop, the most common culprit is the rectifiers.

Disconnect the ac, remove the rectifier diodes from the circuit, and test them. Replace any that are defective.

Apply power again. Still have problems? Disconnect the ac and the rectifiers and apply power again. If the fuse or breaker still pops, the transformer is shorted. Replace it. Reconnect the rectifiers.

Apply power again. If the fuse still blows, disconnect the ac and the filter capacitor. Apply power again.

Problems? If not, the filter capacitor was shorted; replace it. If so, remove power once more and check the regulator components.

Notice that what we are doing is using the process of elimination, starting with the most likely trouble spot. If we disconnect everything from the supply output and the fuse blows, the problem is in the supply. We correct any rectifier problems. Then we check for a bad transformer by disconnecting everything from its output. Next we check for bad filter capacitors by seeing if disconnecting them corrects our problem. Finally, having eliminated everything else, if the trouble remains, we check the regulator. This saves for last the part that is most time consuming to test. It also checks the other parts in order of their likelihood of failure: rectifiers, transformer, and filter. What's more, everything was tested under operating conditions. (*Hint:* Do not memorize the procedure; learn the reasons behind it!)

For power supplies that operate from single-phase 115-V ac, there is a very simple test fixture that can save many fuses. This fixture is shown in Fig. 8-16. It works as follows: When power is applied to a normally functioning power supply, the unit only draws at most one-third of the current required for the bulb to light. This small amount of current is not enough to heat the bulb's filament sufficiently to cause its resistance to rise significantly. (You remember that the resistance of most conductors increases with temperature? Well, a light bulb's resistance increases by a factor of about 4 from its cold value to its value when the bulb is incandescing.) Thus the power supply will receive 80 or 90 V. However, if the power supply is defective and would blow fuses if it could, it will draw all the current that the bulb will let it have. The bulb will limit the current to, at most, three times the normal rated current. The bulb's filament will light up brightly, and very little voltage will be applied to the circuit under test. Thus the bulb serves two purposes. First, it limits the current available to the circuit. The current-limiting action of the bulb is quicker than the action of a fuse. Second, the bulb acts as a fault indicator.

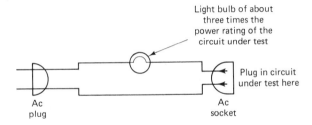

Light bulb of about
three times the
power rating of the
circuit under test

Plug in circuit
under test here

Ac
plug

Ac
socket

Figure 8-16 Light-bulb test fixture.

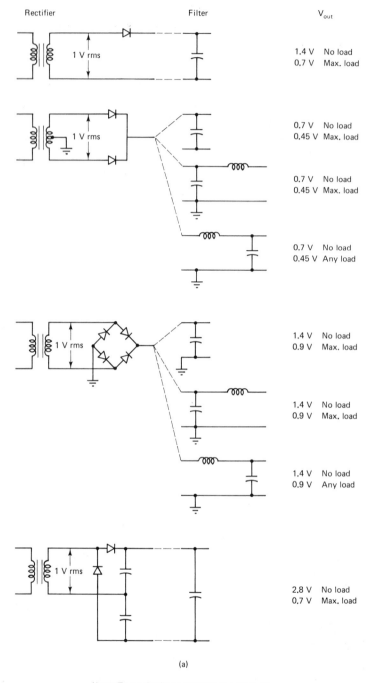

Figure 8-17   Voltage output of rectifier-filter combinations: (a) single-phase rectifiers; (b) three-phase rectifiers.

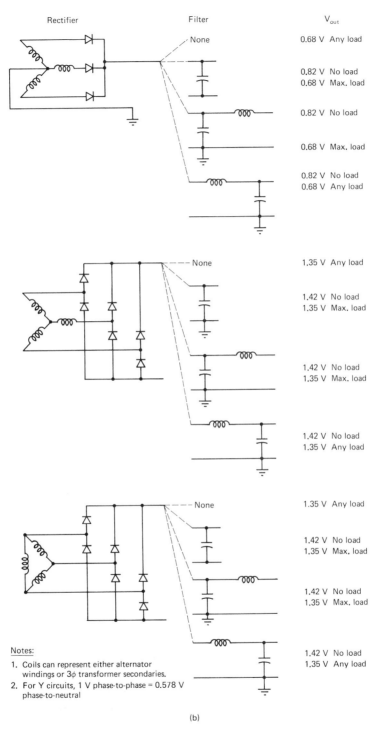

(b)

Figure 8-17 (cont.)

Wrong Output Voltage.   Output voltages, if wrong, are almost always low. (The rare problem of too much output voltage virtually always indicates a defective regulator.) Low output voltage can result from the load drawing too much current, so first disconnect the load. If the voltage is still wrong, the causes, in order of likelihood, are (1) defective regulator, (2) one or more rectifiers defective, (3) filter capacitor leaking dc to ground, or (4) defective transformer. To check for a defective regulator, measure the input and output voltages. If the input voltage is much higher than the (too-low) output voltage, the regulator is probably sick. If they are about the same—within 10% or so—the regulator itself is probably getting too little voltage. In that case, disconnect the ac and test the rectifiers. Replace any defective ones and reapply power. If the output voltage is still low, substitute a filter capacitor known to be good. If the voltage returns to normal, leave the new one in; if not, the old one was good. If all else fails, check the transformer's output voltage to the rectifiers. Too low a voltage indicates a defective transformer. Figure 8-17 should help. Notice that for most filter circuits, there is a difference between no-load (i.e., zero-current) output voltage and full-load (maximum-current) voltage.

No Output Voltage.   We have saved the easiest for last. When there is no output voltage, you just start at the ac supply and check voltages, one stage at a time, until the voltage disappears. The problem will be found at that point.

CHOOSING REPLACEMENT PARTS

Back in Chapter 1, various kinds of electrical and electronic components were introduced, together with their ratings. If you are hazy on this information, now is a good time to review it. I'll wait for you. . . .

Ah, you're back. Now we can summarize what you need to know about choosing replacement parts.

Resistors.   A replacement unit should

1. Have the same value (ohms).
2. Have the same or higher power rating (watts).
3. Have the same or closer tolerance (±%). For example, a 5% resistor can be used to replace a 10% unit, but not vice versa.
4. Be of the same type. The reason for this is that each type has certain advantages over the others. Film resistors are stable, but have poor power-surge handling ability. Carbon-composition resistors have much better surge capacity, but are not as stable. Wirewounds are excellent in both these respects, but introduce inductance at high frequencies. The designer chose the type he wanted.

Capacitors.   A replacement unit should

1. Have the same value (μF or pF).
2. Have the same or higher voltage (WVDC or VAC).
3. Be of the same or better type. For example, an oil-bath or polyester unit can always be substituted for an electrolytic, but the reverse is not true. A polystyrene can be substituted for any other kind. If in doubt, though, stick to the original type. The photos in Chapter 1 should help you to identify the type you are replacing.
4. Be as good or better than the old one with regard to tolerance and/ or temperature range if they are specified. The tolerance is ±20% unless otherwise specified. The maximum temperature is usually 65°C unless otherwise specified. If these are not listed on the old capacitor, assume they do not matter.

Transistors.   Ideally, we would always replace a transistor with a new unit having the same type number. However, in the *real* world, there are discontinued types and many types having only what are called "house numbers." These are type numbers that are meaningful only to the manufacturer of the equipment. Exact replacements, if available at all, must be obtained from this manufacturer, which may be a time-consuming proposition. So the second-best solution is always to look up the defective unit in a semiconductor manufacturer's replacement device handbook. If it is listed, *always* check the ratings of the suggested replacement device to see whether they are adequate. There are some cases in which two very different house-numbered transistors share the same number. Also check the terminal diagram; the replacement unit may have a different lead pattern from the original. In cases in which the original type number is not listed, make your best-educated guess at the necessary characteristics and choose a replacement according to the following guidelines:

1. Same or higher current rating (I_C in mA or amperes).
2. Same or higher breakdown voltage (V_{CEO} in volts; V_{CBO} is usually higher than V_{CEO}, but it is less important in most applications than the V_{CEO}. By the way this means

$$V_{\text{Collector-to-Emitter,base Open}}$$

3. Same or higher power rating (mW or watts)
4. Same polarity (NPN or PNP; or, N-channel or P-channel for FETs)
5. Same material (silicon or germanium)
6. Same or higher gain (h_{fe} or β)
7. Same case. This is preferable, but not essential.

Triacs, SCRs, and Diodes.   The discussion under "transistors" also applies to triacs, SCRs, and diodes. In addition, if a triac circuit does not contain a diac (which looks like a very small diode), suspect that the diac is contained in the triac. This affects both your testing method, as discussed earlier, and your selection of a replacement. If in doubt, a call to the service department of the equipment manufacturer is in order. Otherwise, the replacement unit should have:

1. The same or higher reverse voltage
2. The same or higher forward current
3. Preferably the same type of case

Switches.   A replacement unit should have:

1. The same or higher voltage rating
2. The same or higher contact current rating
3. The same contact configuration (SPST, DPDT, etc.)
4. The same mechanical construction (rotary, slide, toggle, etc.)

Rheostats and Potentiometers.   A replacement unit should have:

1. The same value (ohms)
2. The same or higher power rating
3. Preferably the same mounting provisions

Note: Potentiometers should have the same *taper*, which is the relation of resistance change to shaft rotation. If this is not specified, assume *linear taper*, which means that at 50% rotation you have the wiper at the midpoint of the total resistance.

Coils.   A replacement unit should have:

1. The same inductance (henries)
2. The same or higher current rating

Transformers.   A replacement unit should have:

1. The same primary and secondary voltage ratings
2. The same or higher primary and secondary current ratings
3. The same frequency (usually 60 Hz)

Circuit Breakers.   A replacement unit should have:

1. The same operate current.
2. The same or higher maximum voltage.
3. The same trip time, if specified. (This is the amount of time required for the breaker to open after a specified overcurrent begins to flow.)

SUMMARY

1. Troubleshooting motor systems begins with verifying that there is correct power to the motor.
2. Defective motors should first be checked for mechanical problems. If such problems are found, they should then be isolated to either the motor or the load. Mechanical problems in the motor require that the motor be disassembled.
3. Lack of any response when power is applied may indicate an open or defective thermal circuit breaker ("overload protector" or "overheat protector").
4. Defective polyphase motors having no internal protectors and no mechanical problems usually require rewinding or replacement.
5. Centrifugal switches or current relays are frequent culprits in single-phase motor problems.
6. Motor capacitors, especially electrolytics, sometimes fail.
7. Shaded-pole motors may be disabled by dry or gummy bearings.
8. Faulty windings can sometimes be detected by an ohmmeter; other times, the process of elimination must be used. Repair consists of replacing the armature (if it is a wound type) or rewinding or replacement of the motor.
9. Each of the four types of motor bearings—sleeve, ball, needle, and thrust—has its particular applications. Sleeve bearings are quiet and inexpensive, but subject to wear. Ball bearings have a long life and lower friction, but cost more than sleeve bearings, require more radial space, and are noisier. Needle bearings require less radial space than ball bearings, but are not quite as good in other respects. Thrust bearings are used when the motor's load exerts a force directed along the length of the shaft. Lubrication of bearings is essential, but must be of the proper type, amount, and frequency. Worn bearings can be detected by sound or by excessive side play in the shaft.

10. Gear trains must be kept properly lubricated.

11. Starting switch contacts can often be restored to operation by cleaning.

12. Motor capacitors can be tested with an ohmmeter.

13. Brushes can cause problems as a result of wear, excessive current, oil-soaking, or incorrect spring tension. Defective brushes must be replaced. Rough commutator surfaces should be restored.

14. Specifying a replacement motor requires a knowledge of the original motor type, horsepower, voltage, speed(s), bearing type, shaft style, enclosure style, frame size, and mounting type. Most of this information can be found on the motor's nameplate.

15. Excessive motor noise can be caused by dry or defective bearings or gear trains, by noisy brushes, or by dynamic unbalance.

16. Radiation of noise by a motor can be reduced by means of resilient mountings, proper choice and treatment of mounting surfaces, and/or acoustical enclosures.

17. Electronic components can be tested with an ohmmeter; technicians should be sure they know the proper techniques.

18. The most common component to fail in a speed-control circuit is an SCR or triac.

19. The trouble in a defective power supply can usually be isolated to a particular group of components by proper interpretation of symptoms. Blown fuses indicate excessive current, possibly caused by a defect in the load or by shorted rectifiers, transformer, filter capacitor, or regulator. The same components can cause output voltage to be incorrect.

20. Replacement components must be selected with a knowledge and understanding of the ratings of the original part.

QUESTIONS

1. What is the first step in troubleshooting a motor that is reported to be defective?

2. What two steps are involved in checking a motor for mechanical problems?

3. If a foreign object is found inside a motor, what else should you look for?

4. What can cause a good motor to behave as if its windings were all open?

5. What can be done about a defective polyphase motor whose problem is not mechanical?

6. What can cause a split-phase motor to hum, get hot, and not turn?

7. What can cause a capacitor-start or two-capacitor motor to hum, get hot, and not turn?

8. What can cause a capacitor motor to overheat, even though it runs well?

9. What could cause a two-capacitor motor to have low running torque?

10. What common problem affects shaded-pole motors?

11. How do you test for an open winding?

12. How do you check for a winding shorted to the motor case?

13. Name the four common types of motor bearings and list the advantages and disadvantages of each.

14. What kind of bearing requires virtually no maintenance?

15. What is the proper lubricant for motor bearings? Name two common types of oil that should *not* be used.

16. How can you clean a gummy bearing?

17. How do you test for worn bearings?

18. How should the end-bell mounting screws of a motor be tightened?

19. What can cause noise in gear trains?

20. To what two common defects are mechanical starting switches susceptible?

21. What part is usually at fault if an electronic starting switch fails?

22. Describe the proper method for testing motor capacitors.

23. What are three causes of brush failure?

24. Name four possible causes of excessive brush sparking.

25. How can a slightly scarred commutator be resurfaced?

26. If a split-phase motor has four leads of unidentifiable color, and the ohmmeter readings are as follows, which is the starting winding?

$$\text{Lead 1} - \text{lead 2: 10 } \Omega$$

$$\text{Lead 3} - \text{lead 4:   3 } \Omega$$

27. Name eight ratings that you should know when specifying a replacement motor.

28. What is an explosion-proof motor used for?

29. What is likely to cause a motor to make a raspy, scraping, or squealing sound?

30. What can cause a properly lubricated gear train to be noisy?

31. How can motor vibration noise be reduced?

32. How can midfrequency motor hum and noise be reduced?

33. How can brush noise from a properly operating motor be reduced?

34. Discuss the method of testing each of the following components:
    (a) Resistors
    (b) Capacitors
    (c) Diodes
    (d) Transistors
    (e) SCRs
    (f) Triacs
    (g) Switches

(h) Rheostats or potentiometers

(i) Transformers or chokes

(j) Circuit breakers

(k) Motor starters

35. Why may a triac in a speed-control unit be impossible to test with an ohmmeter?

36. What is the first step in troubleshooting any electronic assembly?

37. What defects can cause a power supply to blow fuses?

38. What defects can cause a power supply to have a low output voltage?

39. What part of a power supply is at fault if the output voltage is too high?

40. If the output voltage of a power supply is low, how do you test the voltage regulator?

41. What is the output voltage of a full-wave bridge power supply feeding a capacitive filter with no regulator if the secondary of the power transformer produces:

(a) 60 V?

(b) 24 V, center-tapped with the center-tap grounded?

42. What is the output voltage of a full-wave power supply (not bridge) with a capacitive filter and no regulator if the power transformer secondary produces 24 V, center-tapped, with the center-tap grounded?

43. How do you troubleshoot a power supply that has no output voltage?

44. Discuss the information needed to specify the following replacement parts:

(a) Resistors

(b) Capacitors

(c) Transistors

(d) Triacs, SCRs, and diodes

(e) Switches

(f) Rheostats and potentiometers

(g) Chokes

(h) Transformers

(i) Circuit breakers

45. Name two possible causes of dynamic unbalance in a motor.

9

LINEAR-MOTION
ELECTROMECHANICAL DEVICES

Most of this book has been devoted to rotating electrical machinery. In Chapter 3, though, we did discuss two types of electromechanical devices whose motion is essentially linear: namely, relays and solenoids. This chapter will deal with more sophisticated linear-motion devices. The two categories of these *linear motors and generators* are (1) linear instrumentation drives and (2) transducers. (A *transducer* is a device that converts information from one form of energy to another. Examples include microphones and loudspeakers.) In discussing these devices, our goal is a bit different from that of the remainder of the book. Repair of linear motors and generators is outside the capabilities of most technicians, primarily because of the specialized equipment involved and the unavailability of parts. However, an understanding of the operation of these devices is necessary for their proper utilization, and essential for intelligent specification of replacement units. Therefore, this chapter will lead you to an understanding of linear motors and generators.

PRINCIPLES OF ELECTROMAGNETIC LINEAR DEVICES

There are many ways of converting electrical energy to mechanical energy, and vice versa. Certainly, the most common is the electromagnetic method. Most electromagnetic linear motors operate more or less like an "uncurled" rotary motor (see Fig. 9-1). The field magnets may be either permanent magnets or electromagnets. In a linear electromagnetic motor, the electrical signal that controls the motion of the coil is fed to the coil leads. The current

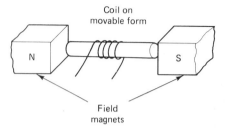

Figure 9-1 Electromagnetic linear
devices—principle of operation.

in the windings of the coil sets up a magnetic field that attracts the coil to
one pole and repels it from the other. Normally, there is some provision for
keeping the coil in a certain "normal" position when no current is applied.
This is usually done by springs. If the device is used as a generator, the motion of coil in the magnetic field results in a voltage being induced in the
coil.

   One example of a linear motor is the pen drive of a chart recorder.
Some of these operate very much like the illustration of Fig. 9-1. A photograph of a typical pen drive motor is shown in Fig. 9-2. The chart recorder

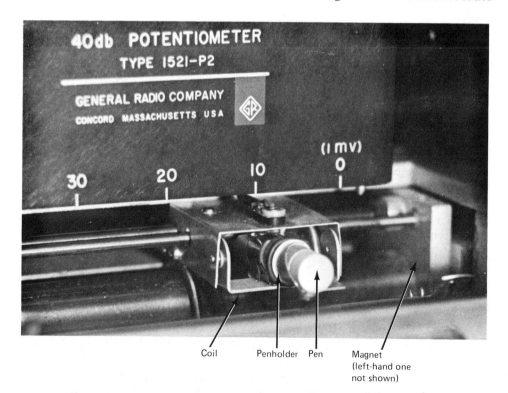

Figure 9-2 Linear motor used in a chart recorder. (Courtesy of Genrad,
Inc.)

draws graphs of voltage versus time. The time axis is provided by moving the paper at a constant speed. The voltage variations are amplified and fed to the pen drive motors. Thus the distance from the pen to the base or zero line depends on the voltage input.

LOUDSPEAKERS

Certainly, the most familiar application of electromagnetic linear motors is the *loudspeaker* or "speaker." Figure 9-3 shows a cross-sectional drawing of a typical loudspeaker. Figure 9-4 shows various means of providing magnetic fields for loudspeakers. In operation, the "lines of force" follow the

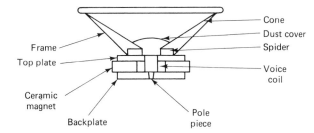

Figure 9-3 Cross section of loudspeaker.

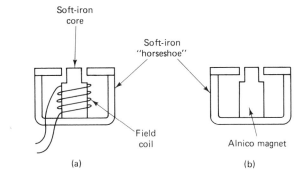

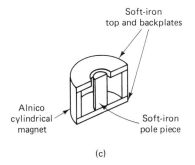

Figure 9-4 Alternate field structures for loudspeakers: (a) electrodynamic; (b) Alnico U structure; (c) Alnico "pot" structure.

paths shown way back in Fig. 2-10, providing a concentrated field around the coil. The *spider* and the edge of the *cone* (the cone edge is also called the *surround* or *skiver*) provide the spring force to keep the *voice coil* centered in the gap.

The primary yardsticks by which a loudspeaker's quality is judged are frequency response, efficiency, and power-handling capability. Let's talk about these one at a time.

The *frequency response* of a loudspeaker is the range of frequencies in which it does a good job of converting electrical energy into acoustical energy. The frequency response required depends on the use to which the speaker is to be put. The human ear responds to vibrations in the range from 20 Hz to about 20,000 Hz. Modern commercial recordings, though, seldom contain information below 60 Hz, and virtually never contain any below 40 Hz. Thus a loudspeaker for optimum reproduction of recorded music would need a low-end response of 60 Hz or below, preferably about 40 Hz. Response below 40 Hz would be unnecessary. Music often has frequency components extending beyond 20,000 Hz. However, tests made by the British Broadcasting Company (BBC) indicate that many people cannot hear any difference when the high-end response of a music system is limited to 12,000 Hz. Less than 1% of their subjects could hear a difference when the high-end response ended at 16,000 Hz. So an adequate loudspeaker for music reproduction would have a response extending from 60 to 12,000 Hz. A very good system would have a response from 40 to 16,000 Hz. For speech-only applications, a much smaller range of frequencies needs to be covered. A response of 300 to 3000 Hz is adequate to provide intelligible speech reproduction. Somewhere in between the speech-only and the music-quality systems are the background-music systems. Such systems are often used in restaurants, department stores, and doctors' offices to provide "acoustical wallpaper." Their frequency response should generally extend from about 100 to 6000 Hz or so.

Between the low-end and high-end cutoff points, the speaker should ideally respond equally to electrical energy at all frequencies. Unfortunately, this ideal is all but impossible to achieve. The more nearly ideal a speaker is in this respect, the better its quality. The frequency response of speaker systems is best described by a graph of acoustical output level versus frequency, with the same power fed to the speaker at all frequencies. Figure 9-5 shows four frequency-response graphs for four different speakers. A very good speaker will have a frequency-response graph that is almost a straight line, not deviating by more than about ±5 dB in between the low- and high-frequency cutoff points. Such a speaker is referred to as having a "flat response." Graph (a) of Fig. 9-5 is the response of a 10-in. woofer, or low-frequency speaker. This unit's response is quite flat from about 50 to 2000 Hz. Graph (b) shows the response of an 8-in. full-range speaker. Notice several things. First, extending the frequency response was achieved at the price of flatness

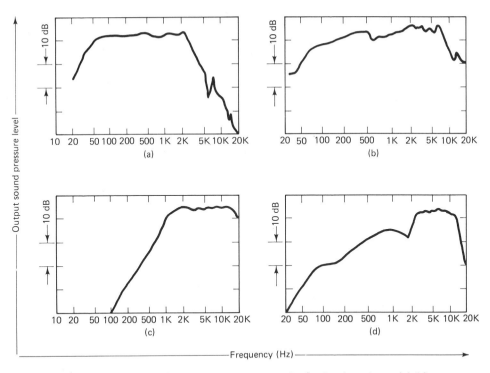

Figure 9-5 Typical frequency-response graphs for loudspeakers: (a) 10-in. woofer; (b) 8-in. full-range speaker; (c) tweeter; (d) 8-in. paging speaker.

of response; that is, even though the response extends from 50 to 10,000 Hz (rather than 50 to 2000, as for the woofer), there is more variation of sensitivity with frequency. The low frequencies are somewhat weaker below about 200 Hz, and there is some raggedness in the response between 500 and 10,000. This speaker would not reproduce music as accurately as one with a flatter response. Graph (c) is the response of a tweeter, or high-frequency speaker. It is quite flat from 1000 to 20,000 Hz. If this tweeter were teamed up in a system with the woofer whose response is shown in graph (a), the result would be a very good system having both wider-range response and flatter response than the full-range speaker of graph (b). Graph (d) shows the response of an 8-in. paging speaker. This speaker's response is intentionally increased in the range 2000 to 10,000 to enhance the articulation, or understandability, of announcements. It would sound "tinny" if used to reproduce music. From this discussion, you can get an idea of the kind of information that can be gained from a speaker's frequency-response graph.

The efficiency of a loudspeaker is not usually expressed in percent. If it were, the figures would not mean much to most people. They would also seem appallingly low: about 0.1 to 24%. The average home stereo speaker converts less than 1% of the electrical input power into sound. Efficiency is usually stated in terms of the sound level (dB) at 1 m from the speaker, when the speaker is powered by 1 W. This is called the *sensitivity* of the speaker. Figure 9-6 is a chart showing sensitivity ranges for various types of speakers. One thing should be kept in mind when you are evaluating speaker sensitivities. The dB figures that are used for measuring sound levels (more correctly, dB SPL—for decibels, sound pressure level) are logarithmically related to power. What this means is that if speaker A has a sensitivity of 93 dB at 1 W at 1 m, it will produce 87 dB at ½ W, 99 dB at 2 W, 102 dB at 4 W, and so on. In other words, twice the power input gives a 3 dB increase in sound level. You can also turn this around: If speaker B produces 90 dB at 1 W at 1 m, it will require 2 W to create the same sound level that speaker A produces with 1 W. In short, 3 dB lower sensitivity implies half the efficiency. Lower efficiency, in turn, means more amplifier power is required.

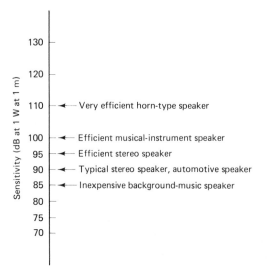

Figure 9-6   Sensitivities of various loudspeakers.

Efficiency depends on many factors, one of which is magnet size. This is the reason that most speaker manufacturers state magnet size as one of a speaker's ratings. Although it is true that a larger magnet often implies a better speaker, there is certainly not a direct relationship. (Magnet weight should not be confused with *magnet structure* weight, which includes the weight of the top plate, back plate, and the pole piece. A 40-oz structure may have a 20-oz magnet.) Manufacturers sometimes exploit public ignorance by selling speakers with larger magnets—and correspondingly larger prices—when a smaller magnet would have served equally well. Table 9-1 should help to clear things up somewhat.

TABLE 9-1 Common Speaker Motor Parameters

Voice coil diameter (in.)	Average magnet size (oz)	Largest useful magnet[a] (oz)	Power-handling capability	
			Standard construction (W)	High-power construction (W)
$1/2$	1	3	$1/2$	2
1	6	16	12	25
$1^1/2$	10–16	30	20	50
2	30–40	54	40	80
$2^1/2$	40–54	80	60	120
3	96	120	—	200
4	96	200	—	200

[a]Ceramic. Alnico magnets are quite expensive and therefore rather rare. A rough comparison is that it takes about one-half the weight of Alnico to produce the same field as a given ceramic magnet. For example, a 16-oz Alnico magnet is roughly equivalent to a 32-oz ceramic. Also, regardless of the trade name used, all types of ferroceramics are roughly equal in performance.

You will notice that we sneaked some information on power-handling capability into Table 9-1. The power-handling ability of a speaker depends primarily on voice-coil size and construction. The coil size is usually given as one of a speaker's ratings. So-called high-power or high-temperature voice coils are wound on aluminum forms rather than the common paper variety. This approximately doubles the power-handling capability of the speaker.

Power-handling capability is specified in watts, continuous. This is the maximum amount of power the speaker can handle for long periods of time. Unfortunately, there is presently no standard rating system to tell how much power a speaker can handle at various frequencies and still sound good. The power rating applies only to the maximum power before thermal destruction occurs. Often, the incorrect term "rms watts" is used. It really means "continuous power," as does "long-term average power." Speakers were once rated in "peak watts"; some still are. This is a useless figure for all practical purposes. The continuous power rating is the only one that should be trusted.

Let's look now at a typical set of specifications for a speaker, in this case for a Soundfair Model X:

Diameter	8 in.
Voice coil	1 in.
Maximum continuous power	12 W
Frequency response	60–10,000 Hz
Sensitivity	92 dB at 1 W at 1 m

Magnet weight	6 oz
Voice coil impedance	8

This speaker would be a pretty good quality background music speaker. Usually, these specifications are available for most replacement speakers. They are usually *not* available for the defective speaker you are replacing. So let's talk about judicious guessing. Standard diameters are 1, 1½, 2, 2½, 3, 3½, 4, 4½, 5, 5½, 6, 6½, 7, 8, 10, 12, 15, and 18 in. There are also elliptical speakers, including mainly 4 in. × 6 in., 4 in. × 10 in., and 6 in. × 9 in. The voice coil diameter can be guessed by looking from the side of the speaker at the apex of the cone, where the cone joins the spider. (Voice coil diameter is not a very important parameter anyway, if the replacement speaker will handle sufficient power.)

The power capability can be assumed equal to the amplifier power divided by the number of full-range speakers in the system. Thus a 10-speaker background music system having a 25-W amplifier must handle 2.5 W per speaker. The necessary frequency response can be estimated from the guidelines given on page 220. The former sensitivity of a dead speaker is essentially unknowable. Usually, though, if you have a choice, choose the replacement speaker having the highest sensitivity. Often there will be choice of sensitivity once the other parameters are matched.

Magnet size can be matched visually.

The voice coil impedance is critical. The replacement speaker's impedance should match that of the original. If the original speaker's voice coil is not open, its coil resistance can be measured with an ohmmeter. The impedance is part inductive reactance, so it is always somewhat higher than the resistance. Thus if a coil measures 3 Ω, it probably has a 4-Ω impedance. The standard impedances are 3.2, 4, 8, 16, and 32 Ω. If the original speaker has an open voice coil, you have several options:

1. In a multispeaker system, measure the impedance of one of the other speakers in the system; they are probably all the same.

2. In a single-speaker system, check the ratings stamped on the back panel of the amplifier that is feeding the speaker.

3. If all else fails, try 8 Ω; that is by far the most common impedance.

All the speaker information that has been given applies equally to any type of speaker. However, speakers used in high-quality sound systems have certain other characteristics that must be matched if the system is to perform properly. Discussing these is a job for a separate book. My intention has been to provide you with a basic understanding of speakers so that you can specify a correct replacement for paging, background music, or communications use.

DYNAMIC MICROPHONES

A dynamic microphone is a linear-motion electromagnetic device that functions as a generator. Its structure is very similar to that of a loudspeaker, as can be seen from Fig. 9-7. The comments about frequency response of loudspeakers also apply to microphones. There are significant differences in the other ratings, however. When selecting a replacement microphone, the most important characteristics besides frequency response are rated impedance and directivity.

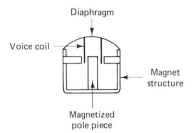

Figure 9-7   Basic structure of a dynamic microphone.

The impedance of most microphone cartridges is around 30 to 50 Ω. Many microphones include a built-in transformer that steps up the impedance to several thousand ohms. The *rated* impedance, though, is much higher. A microphone is designed to operate properly when connected to an impedance that is much higher than the microphone's own impedance. Thus a microphone having an actual impedance of, say, 47 Ω would very likely have a rated impedance of 150 Ω.

The standard rated impedance of microphones are 50, 150, or 250 Ω (low impedance); 600 to 1000 Ω (medium impedance); and about 10,000 Ω (high impedance). A replacement microphone must have the same rated impedance as the original unit. If the impedance is not specified, you can make an educated guess from the type of connector used (see Fig. 9-8).

Some microphones respond almost equally well to sound from any direction; these are called *omnidirectional* microphones. Other microphones are more sensitive to sound coming from the front of the microphone. These are called *unidirectional, directional, cardioid, super-cardioid,* or *hyper-cardioid* microphones. Omnidirectional microphones are the first choice for certain recording applications. For public address, paging, or general sound-reinforcement use, directional microphones are almost always preferred. However, since omnidirectional microphones are cheaper to manufacture, they are often bought for use in applications in which directional microphones would perform better. In other words, when choosing a microphone for any application except recording, choose a directional type.

Oh, by the way, a magnetic stereo cartridge is essentially a dynamic microphone with a stylus instead of a diaphragm.

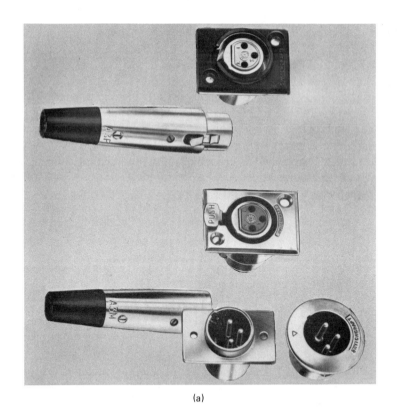

(a)

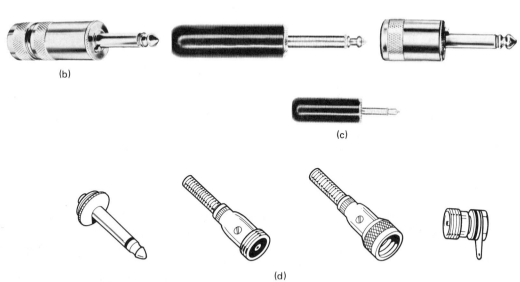

(b)

(c)

(d)

Figure 9-8 Microphone connectors: (a) three-prong connectors (usually low impedance); (b) 1/4-in. phone plug (usually high impedance); (c) 1/8-in miniplug (usually medium impedance); (d) screw-on connectors (usually high impedance). (Courtesy of Switchcraft.)

OTHER TYPES OF ELECTROMECHANICAL DEVICES

There are many principles of operation of electromechanical linear devices beside electromagnetism. We have spent most of our time discussing electromagnetic devices, because they are by far the most common. However, we need to at least introduce the other types.

Piezoelectric Devices

Certain materials—Rochelle salt, quartz, and some ceramics—produce a voltage between the ends of a piece of the material when the material is flexed. Alternatively, if a voltage is applied between two points on one of these materials, the material will flex. This characteristic is called the *piezoelectric* (pie-ee-zo-electric) effect, from the Greek verb *piezon*, "to press." If a piezoelectric material is fastened to a small diaphragm that vibrates in response to sound waves, you have a crystal or ceramic microphone, depending on the type of material (see Fig. 9-9). If it is connected to a large diaphragm and fed by an amplifier, you have a piezoelectric speaker. If it is attached to a steel beam so that it bends when the beam does, you have a sensor whose voltage depends on the amount that the beam bends. This is called a *strain gauge* sensor. If it is mounted on a device that is being vibration-tested, you have an *accelerometer* sensor. If it is attached to a phonograph stylus and tone arm, you have a crystal or ceramic phono cartridge.

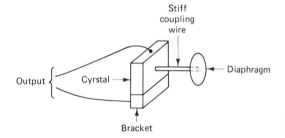

Figure 9-9 Piezoelectric microphone.

Piezoelectric devices are subject to failure primarily from fracture of the crystal or ceramic. This can happen from mechanical causes; or, in piezoelectric speakers, it can happen from the thermal effect of too much input power.

Piezoelectric microphones and speakers have the same types of ratings as electromagnetic ones, except that their impedance is not specified. The impedance of these devices is quite high. Piezoelectric microphones, sensors, and phono cartridges have impedances on the order of a hundred kilohms or so. They require even higher input impedances in order to work properly— 1 MΩ or higher. Piezoelectric speakers have an impedance of around 40 to 80 Ω at 20,000 Hz, and the impedance increases as frequency decreases.

Since they are used primarily as tweeters (high-frequency-only speakers), their impedance is usually ignored; compared to the woofers' (low-frequency speakers) impedance, it looks to an amplifier like an open circuit.

Carbon Microphone

The second microphone invented—and the type that is only gradually being replaced in telephone service—is the carbon microphone. This device is made of carbon granules that are alternately compressed and decompressed. By this means, the contact area of each individual granule with its neighbors is varied. Thus the resistance of the "packet" of carbon is varied in accordance with the vibrations of the sound wave. A dc voltage is applied across the packet of carbon, and the varying resistance modulates it to produce ac; carbon microphones therefore require a dc supply. The sound quality of a carbon microphone is only fair; that's the reason they are becoming less and less common.

Capacitor Transducers

The capacitance of two plates separated by an insulator is given by

$$C = \frac{\epsilon A}{d}$$

where C = capacitance

ϵ = a constant that depends on the dielectric material

A = area of the plate

d = distance between the plates

Also, $C = Q/E$, where Q = quantity of charge and E = polarizing voltage. Now let's do a little algebra;

$$C = \frac{Q}{E} = \frac{\epsilon A}{d}$$

$$Qd = \epsilon A E$$

$$Q = \frac{\epsilon A E}{d}$$

This equation says that if the distance between a capacitor's plates is varied while the voltage is held constant, Q will vary. A varying quantity of charge constitutes a current. If we make the capacitor with one rigid plate and one flexible plate, using air as a dielectric, a sound wave vibrating the flexible plate will cause an audio-frequency current to be generated. This is the principle of the capacitor microphone, also known as the condenser microphone. It is illustrated in Fig. 9-10. In the figure the distance d between the diaphragm and backplate are exaggerated for clarity. Typically, d is about 0.001 in. When the diaphragm vibrates, it varies d. The resistor prevents the battery from shunting out the microphone's ac output voltage. Since the microphone draws no direct current, the dc drop across the resistor is zero volts, and the full battery voltage is available for polarizing.

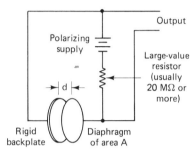

Figure 9-10 Capacitor microphone.

Capacitor microphones require extremely high impedance amplifier inputs, so they are often built with integrated-circuit preamplifiers inside the microphone case. Traditional capacitor microphones also require a polarizing voltage. However, in the late 1960s, researchers developed the *electret capacitor microphone*. This is a device using a thin, metallized plastic diaphragm with a built-in electric field. Electret microphones therefore do not require a polarizing voltage, although they do require from 1½ to 9 V to supply the internal preamplifier.

If a capacitor microphone is fed by an ac voltage, it will produce sound. Then you have an electrostatic speaker. These are not particularly common, although several companies do produce them. A few companies produce full-range electrostatic speakers, but electrostatics are most often used as tweeters. While the electret principle is applicable to headphones, most electrostatic speakers use a separate polarizing supply.

Magnetostriction Devices

"Magneto-what?" Well, it's like this. When certain materials are magnetized suddenly, their physical dimensions change. When they are placed in a magnetic field and they are mechanically stressed, their magnetization changes. This is called the magnetostrictive effect. Magnetostriction transducers can be used either to convert electrical energy to acoustical energy, or vice versa. They have a low impedance, which makes them useful as strain-gauge sensors. They also find use as sonar transmitters and receivers.

Ribbon Transducers

If the coil of a dynamic microphone were reduced to one turn and rolled flat, you would have a ribbon microphone. This device consists of a thin, corrugated aluminum ribbon suspended in a strong magnetic field (see Fig. 9-11). Ribbon microphones have much less diaphragm mass than dynamic microphones, enabling them to respond a bit more accurately to complex sounds. Their extremely low output impedance is usually stepped up to some standard value via a built-in transformer. In many applications, ribbon microphones have been replaced by capacitor microphones; their diaphragms also have low mass and are physically more rugged. The ribbon microphone principle is reversible; that is, ribbon speakers are not only possible, they are found in some commercial stereo systems. They are limited to tweeter applications.

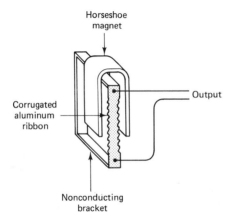

Figure 9-11   Ribbon microphone.

SUMMARY

1. Electromagnetic linear devices operate by means of a coil suspended in a magnetic field. Motion of the coil generates a voltage, making a microphone or phonograph cartridge. If a coil is driven by a current, it will move in the field and the electromagnetic linear motor can be used as a part of a speaker or instrument drive.

2. Speakers are rated in terms of size, frequency response, sensitivity, power-handling capability, and voice coil impedance.

3. A speaker's sensitivity is related to efficiency and is partly determined by a magnet weight.

4. A speaker's power-handling capability is determined by voice coil size and construction.

5. Dynamic microphones utilize the electromagnetic principle. Like other types of microphones, they are rated in terms of frequency response, directivity, and impedance.

6. Piezoelectric devices incorporate a material that produces a voltage when flexed, or flexes when a voltage is applied. They may be used as microphones, speakers, or various kinds of sensors.

7. Carbon microphones turn dc into ac by the variation of their resistance in response to a sound wave.

8. Capacitor microphones and speakers use the dependence of capacitive charge upon voltage and spacing to convert sound into an electrical signal, or vice versa. Traditional ones require a polarizing voltage; electret types do not. Capacitor microphones employ internal preamplifiers to step down the extremely high impedance of the microphone element.

9. Magnetostriction devices use the interdependence of size and magnetization of certain materials. They can function as speakers, microphones, or sensors.

10. Ribbon microphones and speakers use the electromagnetic effect. The element consists of a thin conducting ribbon suspended in a magnetic field.

QUESTIONS

1. Describe the operation of a loudspeaker.
2. What three basic kinds of magnets can be used in the field structure of a speaker?
3. How does a linear electromagnetic pen drive for a chart recorder work?
4. What frequency response should a microphone or speaker have if it is to be used for:
   (a) High-fidelity music reproduction?
   (b) Paging?
   (c) Background music?
5. How is the sensitivity of a loudspeaker expressed?
6. If a speaker whose sensitivity is 96 dB at 1 W at 1 m is replaced by one whose sensitivity is 108 dB at 1 W at 1 m, by what factor can the amplifier's power be reduced?
7. Is a speaker with a 20-oz magnet always better than one with a 10-oz magnet? Why or why not?
8. How much power can a standard 8-in speaker with a 1-in. voice coil probably handle?
9. Name three ways to approximate the voice-coil impedance of a speaker.
10. If a defective speaker measures $10^{1}/4$ in. in diameter, what standard replacement size would you specify?
11. If a paging system uses seven speakers and the amplifier power is 27 W, what is the minimum power rating for a replacement unit?

12. True or false: The rated impedance of a microphone can be determined with an ohmmeter.

13. What are the standard impedances for:
    (a) Low-impedance microphones?
    (b) Medium-impedance microphones?
    (c) High-impedance microphones?

14. What is the typical rated impedance of a piezoelectric microphone?

15. What are two other names for a piezoelectric phono cartridge?

16. What is the difference between a cardioid and an omnidirectional microphone? Which is preferable for PA?

17. How does a piezoelectric speaker work?

18. How does a carbon microphone work?

19. How does a capacitor microphone work?

20. What is the advantage of an electret element over a traditional capacitor microphone element?

21. Why is a battery used in an electret capacitor microphone?

22. Describe the magnetostrictive effect.

23. Name two applications of magnetostriction transducers.

24. Describe the operation of a ribbon microphone.

25. How does a ribbon speaker work?

A

METRIC UNITS
AND CONVERSIONS

METRIC SYSTEM CONVERSION FACTORS

Length

Centimeter	=	0.3937 inch
Meter	=	3.28 feet
Meter	=	1.094 yards
Kilometer	=	0.621 statute mile
Kilometer	=	0.5400 nautical mile
Inch	=	2.54 centimeters
Foot	=	0.3048 meter
Yard	=	0.9144 meter
Statute mile	=	1.61 kilometers
Nautical mile	=	1.852 kilometers

Area

Sq centimeter	=	0.155 sq inch
Sq meter	=	10.76 sq feet
Sq meter	=	1.196 sq yards
Hectare	=	2.47 acres
Sq kilometer	=	0.386 sq mile
Sq inch	=	6.45 sq centimeters
Sq foot	=	0.0929 sq meter
Sq yard	=	0.836 sq meter
Acre	=	0.405 hectare
Sq mile	=	2.59 sq kilometers

Volume

Cu centimeter	=	0.0610 cu inch
Cu meter	=	35.3 cu feet
Cu meter	=	1.308 cu yards
Cu inch	=	16.39 cu centimeters
Cu foot	=	0.0283 cu meter
Cu yard	=	0.765 cu meter

Capacity

Milliliter	=	0.0338 U.S. fluid ounce
Liter	=	1.057 U.S. liq quarts
Liter	=	0.908 U.S. dry quart
U.S. fluid ounce	=	29.57 milliliters
U.S. liq quart	=	0.946 liter
U.S. dry quart	=	1.101 liters

Mass or Weight

Gram	=	15.43 grains
Gram	=	0.0353 avdp ounce
Kilogram	=	2.205 avdp pounds
Metric ton	=	1.102 short or net tons

Source: Reprinted from *Electronics Vest-Pocket Reference Book*, by Harry Thomas, reprinted courtesy of Prentice-Hall, Inc. © 1969 by Harry Thomas.

Grain   =   0.0648 gram
Avdp ounce   = 28.35 grams
Avdp pound   =   0.4536 kilogram
Short or net ton =   0.907 metric ton

Multiples and Submultiples	Prefixes	Symbols
$1,000,000,000,000 = 10^{12}$	tera	T
$1,000,000,000 = 10^{9}$	giga	G
$1,000,000 = 10^{6}$	mega	M
$1,000 = 10^{3}$	kilo	k
$100 = 10^{2}$	hecto	h

$10 = 10$	deka	dk
$0.1 = 10^{-1}$	deci	d
$0.01 = 10^{-2}$	centi	c
$0.001 = 10^{-3}$	milli	m
$0.000001 = 10^{-6}$	micro	μ*
$0.000000001 = 10^{-9}$	nano	n
$0.000000000001 = 10^{-12}$	pico	p

EXAMPLE: 1000 meters (or 10^3 meters) is called a kilometer, and one millionth of a gram (or 10^{-6} gram) is called a microgram.

*1 millionth of a meter is called a *micron*, and is abbreviated simply μ.

B

QUANTITATIVE ANALYSIS OF SERIES AND PARALLEL AC CIRCUITS

Ac circuit analysis can be done using either of two systems of notation. The system used in Chapter 3—phasors having magnitude and phase angle—is called *polar notation.* The other system is called *rectangular or complex* notation. Both systems have advantages. Let's look at the impedance of a series *RCL* circuit:

$$R = 10 \ \Omega$$

$$X_C = 20 \ \Omega$$

$$X_L = 30 \ \Omega$$

In polar notation we would find the impedance as follows:

$$Z = \sqrt{R^2 + (X_L - X_C)^2} \quad \underline{/\tan^{-1}\left(\frac{X_L - X_C}{R}\right)}$$

$$= \sqrt{(10)^2 + (30 - 20)^2} \quad \underline{/\tan^{-1}\left(\frac{30 - 20}{10}\right)}$$

$$= 200 \ \underline{/\tan^{-1}(1)} = 14.14 \Omega \ \underline{/45^\circ}$$

In rectangular notation, we use what is called the *j* operator; it represents an imaginary number whose square equals −1. More importantly for us, it serves as a label to keep resistance and reactance separate. To find the impedance of the same circuit in rectangular notation, we calculate as follows:

$$Z = R + j(X_L - X_C)$$
$$= 10 + j(30 - 20)$$
$$= 10 + j10 \ \Omega$$

Notice that the first number is the resistance, and the j factor ($j \ 10 \ \Omega$) is just the net reactance. In other words, an impedance in rectangular form is j times the net reactance ($X_L - X_C$), added to the resistance. No angle is specified. Another way of looking at all this is that polar notation specifies the hypotenuse and the angle of the impedance triangle, while rectangular notation specifies the other two sides (opposite and adjacent).

As you would expect, there is a very simple system for converting from polar to rectangular, and vice versa:

$$Z_{\text{rect}} = |Z|_{\text{polar}} \cos \theta + j(|Z|_{\text{polar}} \sin \theta)$$

Thus in our example,

$$Z_{\text{rect}} = 14.14 \ \Omega \cos 45° + j(14.14 \ \Omega \sin 45°)$$

Both sine and cosine of $45°$ equal 0.707; we insert them into the equation:

$$Z_{\text{rect}} = 14.14 \ \Omega \times 0.707 + j(14.14 \ \Omega \times 0.707)$$
$$= 10 + j10 \ \Omega$$

which is just what we would expect. The very process of calculating the polar-form impedance is a rectangular-to-polar conversion.

Now, the interesting thing is that it is easy to multiply and divide impedances in the polar form, but difficult to add or subtract them. It is easy to add and subtract impedances in the rectangular form, but difficult to multiply or divide them:

$$Z_{1\,\text{polar}} \times Z_{2\,\text{polar}} = |Z_1| \times |Z_2| \underline{/\theta_1 + \theta_2}$$

$$\frac{Z_{1\,\text{polar}}}{Z_{2\,\text{polar}}} = \frac{|Z_1|}{|Z_2|} \underline{/\theta_1 - \theta_2}$$

For example, let

$$Z_1 = 10 \ \Omega \underline{/15°}$$
$$Z_2 = 43 \ \Omega \underline{/9°}$$

Then

$$Z_1 \times Z_2 = 10 \times 43 \ \Omega \underline{/(15° + 9°)} = 430 \ \Omega \underline{/24°}$$

and

$$\frac{Z_1}{Z_2} = \frac{10 \ \Omega}{43 \ \Omega} \underline{/(15° - 9°)} = 0.23 \ \Omega \underline{/6°}$$

Also:

$$Z_{1\,\text{rect}} + Z_{2\,\text{rect}} = R_1 + R_2 + j(X_1 + X_2)$$

$$Z_{1\,\text{rect}} - Z_{2\,\text{rect}} = R_1 - R_2 + j(X_1 - X_2)$$

For example, let

$$Z_1 = 40 + j22\ \Omega$$

$$Z_2 = 10 - j10\ \Omega*$$

Then

$$Z_1 + Z_2 = 40 + 10 + j[22 + (-10)] = 50 + j12\ \Omega$$

and

$$Z_1 - Z_2 = 40 - 10 + j[22 - (-10)] = 30 + j32\ \Omega$$

Why do we want to do all this? In order to be able to calculate parallel circuits. For example, let's solve for the impedance of the circuit in Fig. B-1.

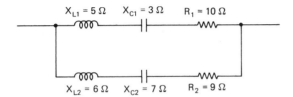

Figure B-1

If the impedance of the top branch is Z_1 and the impedance of the bottom branch is Z_2, the total impedance is given by either:

$$Z_T = \frac{Z_1 Z_2}{Z_1 + Z_2}$$

or

$$Z_T = \frac{1}{\dfrac{1}{Z_1} + \dfrac{1}{Z_2}}$$

(Remember the parallel resistance formulas?) Notice that we have to multiply, divide, and add to solve these equations. Let's see how it is done, using the upper formula. First, find Z_1 and Z_2 in both forms:

$$Z_1 = 10 + j(5 - 3) = 10 + j2 = 10.2\ \Omega\underline{/11^\circ}$$

$$Z_2 = 9 + j(6 - 7) = 9 - j1 = 9.06\ \Omega\underline{/-6.5^\circ}$$

---

*A negative j factor indicates a capacitive reactance.

Second, solve for the product:

$$Z_1 Z_2 = (10.2 \underline{/11°}) \times (9.06 \underline{/-6.5°}) = 92.4 \; \Omega \underline{/4.5°}$$

Third, solve for the sum:

$$Z_1 + Z_2 = 10 + 9 + j[2 + (-1)]$$
$$= 19 + j1 = 19.03 \; \Omega \underline{/3°}$$

Finally, divide the product by the sum:

$$\frac{Z_1 Z_2}{Z_1 + Z_2} = \frac{92.4 \; \Omega \underline{/4.5°}}{19.03 \; \Omega \underline{/3°}} = 4.86 \; \Omega \underline{/1.5°}$$

Take my word for it, you do not want to try this using the lower Z_T formula on page 237 unless you own stock in a paper company! In fact, if your circuit has more than two branches, you are better off solving for two at a time using the approach described above.

If you have several parallel circuits in series, you just convert your answers back into rectangular form and add them all up.

APPENDIX

C

TABLE OF TRIGONOMETRIC FUNCTIONS

TABLE OF SINES, COSINES AND TANGENTS

Angle	Radians	Sine	Cosine	Tangent
0°	.0000	.0000	1.0000	.0000
1	.0175	.0175	.9998	.0175
2	.0349	.0349	.9994	.0349
3	.0524	.0523	.9986	.0524
4	.0698	.0698	.9976	.0699
5	.0873	.0872	.9962	.0875
6	.1047	.1045	.9945	.1051
7	.1222	.1219	.9925	.1228
8	.1396	.1392	.9903	.1405
9	.1571	.1564	.9877	.1584
10	.1745	.1736	.9848	.1763
11	.1920	.1908	.9816	.1944
12	.2094	.2079	.9781	.2126
13	.2269	.2250	.9744	.2309
14	.2443	.2419	.9703	.2493
15	.2618	.2588	.9659	.2679
16	.2793	.2756	.9613	.2867
17	.2967	.2924	.9563	.3057
18	.3142	.3090	.9511	.3249
19	.3316	.3256	.9455	.3443
20	.3491	.3420	.9397	.3640

Angle	Radians	Sine	Cosine	Tangent
21	.3665	.3584	.9336	.3839
22	.3840	.3746	.9272	.4040
23	.4014	.3907	.9205	.4245
24	.4189	.4067	.9135	.4452
25	.4363	.4226	.9063	.4663
26	.4538	.4384	.8988	.4877
27	.4712	.4540	.8910	.5095
28	.4887	.4695	.8829	.5317
29	.5061	.4848	.8746	.5543
30	.5236	.5000	.8660	.5774
31	.5411	.5150	.8572	.6009
32	.5585	.5299	.8480	.6249
33	.5760	.5446	.8387	.6494
34	.5934	.5592	.8290	.6745
35	.6109	.5736	.8192	.7002
36	.6283	.5878	.8090	.7265
37	.6458	.6018	.7986	.7536
38	.6632	.6157	.7880	.7813
39	.6807	.6293	.7771	.8098
40	.6981	.6428	.7660	.8391
41	.7156	.6561	.7547	.8693
42	.7330	.6691	.7431	.9004
43	.7505	.6820	.7314	.9325
44	.7679	.6947	.7193	.9657
45	.7854	.7071	.7071	1.0000

(Continued on next page)

Source: Reprinted from *Electronics Vest-Pocket Reference Book,* by Harry Thomas, reprinted courtesy of Prentice-Hall, Inc. © 1969 by Harry Thomas.

Angle	Radians	Sine	Cosine	Tangent	Angle	Radians	Sine	Cosine	Tangent
46	.8029	.7193	.6947	1.0355	71	1.2392	.9455	.3256	2.9042
47	.8203	.7314	.6820	1.0724	72	1.2566	.9511	.3090	3.0777
48	.8378	.7431	.6691	1.1106	73	1.2741	.9563	.2924	3.2709
49	.8552	.7547	.6561	1.1504	74	1.2915	.9613	.2756	3.4874
50	.8727	.7660	.6428	1.1918	75	1.3090	.9659	.2588	3.7321
51	.8901	.7771	.6293	1.2349	76	1.3265	.9703	.2419	4.0108
52	.9076	.7880	.6157	1.2799	77	1.3439	.9744	.2250	4.3315
53	.9250	.7986	.6018	1.3270	78	1.3614	.9781	.2079	4.7046
54	.9425	.8090	.5878	1.3764	79	1.3788	.9816	.1908	5.1446
55	.9599	.8192	.5736	1.4281	80	1.3963	.9848	.1736	5.6713
56	.9774	.8290	.5592	1.4826	81	1.4137	.9877	.1564	6.3138
57	.9948	.8387	.5446	1.5399	82	1.4312	.9903	.1392	7.1154
58	1.0123	.8480	.5299	1.6003	83	1.4486	.9925	.1219	8.1443
59	1.0297	.8572	.5150	1.6643	84	1.4661	.9945	.1045	9.5144
60	1.0472	.8660	.5000	1.7321	85	1.4835	.9962	.0872	11.43
61	1.0647	.8746	.4848	1.8040	86	1.5010	.9976	.0698	14.30
62	1.0821	.8829	.4695	1.8807	87	1.5184	.9986	.0523	19.08
63	1.0996	.8910	.4540	1.9626	88	1.5359	.9994	.0349	28.64
64	1.1170	.8988	.4384	2.0503	89	1.5533	.9998	.0175	57.29
65	1.1345	.9063	.4226	2.1445					
66	1.1519	.9135	.4067	2.2460					
67	1.1694	.9205	.3907	2.3559					
68	1.1868	.9272	.3746	2.4751					
69	1.2043	.9336	.3584	2.6051					
70	1.2217	.9397	.3420	2.7475					

D

ALTERNATE
SCHEMATIC SYMBOLS

Schematic Wiring Diagram Symbols				
Component	Function	Part	Symbol	Ltr.
Circuit breaker	Opens or closes circuit by nonautomatic means and automatically opens circuit at a predetermined overload current	Air-insulated contacts		CB
		Oil-insulated contacts	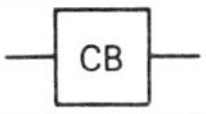	
Contactor	Makes and breaks power circuit to the load when coil is energized or de-energized	Coil	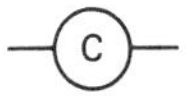	C
		Normally open contact	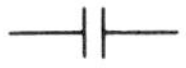	
		Normally closed contact		
Control relay	Energizes or de-energizes electrically operated devices when coil is energized or de-energized	Coil	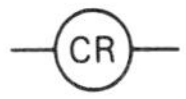	CR
		Normally open contact	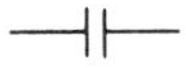	
		Normally closed contact		
Current transformer	Induces low current in secondary wound around primary	Winding and primary		CT
Current transformer with ammeter	To record induced current in secondary	Winding, primary ammeter and ground	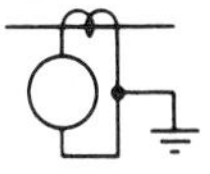	CT
Float switch	Makes or breaks when actuated by float or other liquid-level device	Float and normally open contact		FS
		Float and normally closed contact		

Source: Reprinted by permission of Prentice-Hall, Inc., from *Electrical Drafting*, by Errol G. Schreiver. © 1984.

Schematic Wiring Diagram Symbols				
Component	Function	Part	Symbol	Ltr.
Foot switch	Makes or breaks contacts when manually activated	Normally open contact		FTS
		Normally closed contact		
Fuse	Breaks circuit when predetermined current melts conducting element	Fuse body	* Rating	FU
Knife switch	Makes or breaks circuit when manually engaged or disengaged	Contacts		KS
Motor starter	Makes or breaks power circuit to motor when coil is energized or de-energized	Coil overload relay and contacts	M	M
		Normally closed contact		
Overload relay	Disconnects motor starter at predetermined overcurrent	Magnetic coil		OL
		Thermal element		
		Contacts		
Pilot light	Visually indicates presence of voltage in a circuit (color is indicated by initial)	Lamp		PL

Schematic Wiring Diagram Symbols				
Component	Function	Part	Symbol	Ltr.
Voltage transformer	Reduces or increases voltage when current in primary windings induces voltage in secondary windings	Windings		XFMR
Ground	Connects conductor or device to ground	Grounding connection to ground		GRND
Terminal	Internal terminal in motor starter to which connections are made	Terminal * = terminal number	*	T
Solenoid valve	Electrically operated air valve, which in turn actuates a larger device	Valve and actuator		SOL
Resistor	Reduces voltage in a circuit through introduction of resistance	Resistor (fixed value)	R	R
Rectifier	Converts alternating current to direct current	Full-wave rectifier		REC

Source: Both of the above reprinted by permission of Prentice-Hall, Inc., from *Electrical Drafting*, by Errol G. Schreiver. © 1984.

Schematic Wiring Diagram Symbols				
Component	Function	Part	Symbol	Ltr.
Push to test pilot light	Combines pilot light with self-testing circuit and switch	Lamp and switch		
Pressure switch	Makes or breaks circuit when pressure or vacuum bellows actuates switch contacts	Normally open contact (NO)		PS
		Normally closed contact (NC)		
Push-button switch	Makes or breaks circuit when plunger is manually depressed or released	Normally open contact (NO)		PB
		Normally closed contact (NC)		
		Combined NO and NC contacts		
Rheostat	Reduces voltage in a circuit through resistance winding and tapping the reduced voltage from a sliding contact	Terminals and slider		RH
Temperature switch	Makes or breaks circuit when a pre-determined temperature is attained	Normally open contact (NO)		TS
		Normally closed contact (NC)		
Terminal	Provides connection point	Tie-point or splice (connection)		

Schematic Wiring Diagram Symbols				
Component	Function	Part	Symbol	Ltr.
Timing relay	On: Delay retards relay-contact action for predetermined time after coil is energized	Normally open timed closing (NOTC) contact		TR OR TDR
		Normally closed timed opening (NCTO) contact		
		Normally open timed open (NOTO) contact		
	Off: Delay retards relay-contact action for predetermined time after coil is de-energized	Normally closed timed closed (NOTC) contact		
		No instantaneous contact		
		No instantaneous contact		
Limit switch	Makes or breaks control circuit when mechanically actuated by motion or part of a powered machine	Normally open contact		LS
		Normally open contact held closed		
		Normally closed contact		
		Normally closed held open contact		

Source: Both of the above reprinted by permission of Prentice-Hall, Inc., from *Electrical Drafting*, by Errol G. Schreiver. © 1984.

Schematic Wiring Diagram Symbols				
Component	Function	Part	Symbol	Ltr.
Space heater	To provide heat to cubicle or motor during periods of de-energized state to prevent moisture accumulation.	Space heater filament	—ᴡᴡᴡ—	
Mushroom head pushbutton	To provide exaggerated surface on pushbutton actuator; used primarily as emergency switches	Head and plunger	—o⊥o—	PB
Solenoid valve	Magnetic plunger, electrically actuated	Solenoid	—o⌐o—	SOL
Flow switch	Makes or breaks circuit when actuated by predetermined flow or no-flow status	Normally open contact		FLS
		Normally closed contact		
Hand-off automatic switch	Provides contacts to manually energize a motor or device and bypass permissive devices or to allow control devices to close contacts in auto position	Switch and contacts in off position		HOA
Capacitor	(Condensor) A device consisting of two electrodes separated by a dielectric, which may be air, for introducing capacitance into a circuit	Electrodes		CAP

Source: Reprinted by permission of Prentice-Hall, Inc., from *Electrical Drafting,* by Errol G. Schreiver. © 1984.

INDEX

S

T